Study Guide for

Principles of Statistics

Elke U. Weber
Harvard University

Krieger Publishing Company
Malabar, Florida

Original Edition 1983
Reprint Edition 1989

Printed and Published by
ROBERT E. KRIEGER PUBLISHING COMPANY, INC.
KRIEGER DRIVE
MALABAR, FLORIDA 32950

Printed in the United States of America

Library of Congress Cataloging-in-Publication Data

ISBN 0-89464-409-2

10 9 8 7 6 5 4 3

TABLE OF CONTENTS

Acknowledgements

Throughout the planning and development of this study guide, Paul Herzberg took an active interest in making it a truly integrated part of his statistics package.

I am grateful to Linda Di Francesco for writing Chapter 1, on math preparation, and for her assistance in developing parallel quiz forms for some of the other chapters.

Grant Austin made most of the final revisions to the study guide. The camera-ready copy was prepared by Doris Rippington and Rhoda Nogiec.

TO THE INSTRUCTOR

This study guide differs from conventional study guides in perhaps more ways than are immediately apparent. It was developed in close conjunction with Dr. Herzberg's text and with four sets of multiple-choice quizzes for each chapter of the text, making it an integral part of a three-part teaching package. A description of the genesis of study guide and quizzes will put their pedagogic features into context.

The idea for the format of the study guide grew out of my attempts to produce a "nice" set of multiple-choice test-items for an introductory statistics course in the summer of 1979. Designing multiple-choice items with a sufficient number of satisfactory alternative answers is no easy task. Educational literature tries to assist, advising for example that "*misleads* should be *seductive* to those (students) with partial information or faulty thinking" (Ayers, 1967). The test constructor is to ask himself whether "*this distractor looks good enough* for some students to choose it" (Ebel, 1972). While these suggestions offer little practical advice on how to generate alternative answers, the choice of words is perhaps indicative of a prevalent attitude towards multiple-choice tests. "Distractor" or "mislead" (rather than "alternative") conveys a sense of trickery or deceit which, indeed, is often voiced by students. To describe the desirable attributes of such misleads as "seductive" or "looking good enough", assigns test-construction to intuition or, at best, to a trial-and-error procedure.

TO THE INSTRUCTOR

Thus the designer of multiple-choice tests is left with the problem of developing a methodology that will produce acceptable items within a reasonable time-period. After administering the test, the instructor is faced with the difficulty of assessing the nature of the misconception(s) underlying incorrect student-choices without being able to refer to accompanying work. As a result, students are often left without diagnostic and remedial feedback on their performance.

A systematic approach to test-design is rewarded by providing test-items that tell more than whether a student is "right or wrong". Ideally, each answer alternative to the correct one can signal the presence of a specific misconception. Thus, beside its evaluative function, the test can serve as an easily scored diagnostic tool.

Two major steps are involved in the development of such tests. The first one, namely the categorization of behavioral objectives of the instructional unit to be tested, is in no way unique to the development of multiple-choice items. Crossing major content areas with appropriate cognitive objectives, the instructor creates a hierarchy of specific behaviors which the student should be capable of exhibiting subsequent to his/her exposure to the program of instruction.

The advantages of such a taxonomy for the development of any type of test are obvious. While providing a check on the representativeness of a test, the taxonomy also enables the instructor to build a controlled bias into the test if he/she so desires. In addition, being a rough model of the knowledge base underlying an instructional unit, the taxonomy serves as a stepping stone to the second stage of this test-design procedure, i.e., the development of diagnostic models of faulty behavior.

I initially employed Bloom's (Bloom, Engelhart, Hill, First, & Krathwohl, 1956) taxonomy of educational objectives to classify the contents of each text chapter, but, in the process, Bloom's six cognitive categories (knowledge, comprehension, application, analysis, synthesis, and evaluation) were condensed to three: knowledge, comprehension, and application. At the introductory level, a statistics course places little or no emphasis on evaluation or even synthesis--perhaps deplorably so. There also seemed to be little practical difference between the levels of application and analysis. Thus the two were collapsed into a single category, application, which also includes elements of analysis and synthesis as defined by Bloom. Other educators suggest somewhat similar three- or four-stage models (Forsyth, 1977; Medcof, 1978).

Both the practice quiz and the four sets of actual quizzes (which are identical in format and parallel in content) were biased in favor of the two higher levels of cognitive functioning: comprehension and application. This is in contrast to most conventional multiple-choice items which often test at little more than the knowledge level.

Diagnostic models are "models that capture a student's common misconceptions or faulty behavior as simple changes to the correct model of the underlying knowledge base" (Brown & Burton, 1978). As already mentioned, the behavior-content matrix can be seen as a large-scale model of interrelationships between knowledge components ("knowledge" understood in a more general sense than defined by Bloom). Yet, behavior-objectives stated in the matrix are often not terminal, i.e., can themselves be reduced into several conceptual and/or operational components. An example will illustrate this point.

Take the task of determining the value of β for the $\underline{z}$-test of the mean of one population, given that two specific hypotheses $\underline{H}_o$: $\mu = \mu_o$, $\underline{H}_1$: $\mu = \mu_1$ and the critical value $\bar{\underline{y}}_{crit}$ are known. Let us assume that $\beta = \text{Prob}(\bar{\underline{y}}_{obs} < \bar{\underline{y}}_{crit})$ under $\underline{H}_1$: $\mu = \mu_1$. A student has to make a series of correct decisions to arrive at this answer. He/She has to decide whether to a) use μ_o or μ_1 in the calculation of $\underline{z}_{crit}$, b) apply the central limit theorem (σ or $\sigma/\sqrt{\underline{n}}$ in the denominator), c) place the region of β to the left or right of $\underline{z}_{crit}$ [$\text{Prob}(\underline{z} < \underline{z}_{crit})$ or $\text{Prob}(\underline{z} > \underline{z}_{crit})$], and d) add or subtract 0.5 to the probability obtained from the $\underline{z}$-table or leave it (as decided from a proper diagram). If represented in a tree diagram, these four consecutive decisions lead to sixteen potential outcomes, each outcome being a unique combination of correct and/or faulty decisions. Employing these outcomes as alternative answers of the test-item, the instructor is able to diagnose any incorrect answer as a specific set of misconceptions.

Unfortunately, several factors complicate this procedure. The assumption, that all outcomes of the decision tree are distinct outcomes, is not always warranted. In some cases, two incorrect steps may accidentally cancel one another, leaving the final (reported) answer indistinguishable from the correct answer. Also, a major conceptual error may have an only minimal numerical manifestation which could go unnoticed after appropriate rounding. In these situations additional test-items are needed to discover which combination of decisions underlies the student's answer.

Also basic to the procedure is the premise that a student's choice is based on reasoning rather than guessing. Without this premise diagnosis becomes meaningless. Subtracting points for incorrect answers can help minimize the frequency of guessing.

For tasks of some complexity (e.g. determining the value of β) not all possible variations of the task-solving process can practically be included into the item-design. With an optimal number of five or six alternatives for multiple-choice items, selection is necessary. The obvious way to determine which combinations of misconceptions are most prevalent is to consult written answers produced by students on conventional (completion form) tests. These answers show the strategy used by the student as he/she proceeds in solving the given problem.

I had approximately 250 such quizzes for each instructional unit (text chapter), written in previous years by students of the course. An error-analysis of these quizzes reduced the number of alternatives for any given task considerably. For the β-value example only four of the possible sixteen outcomes were, in fact, observed. In some cases, the quiz-analysis added branches to the theoretical diagnostic models. E.g., when calculating β, many students stopped halfway through the task, reporting the (correct) value of $\underline{z}_{crit}$ incorrectly as β. Thus, an error analysis of existing tests in completion form (student-produced answers) can be instrumental in validating the task-decomposition of theoretical diagnostic models.

TO THE INSTRUCTOR

Bloom (1971) observed that "students respond best to diagnostic results when the diagnosis is accompanied by a very specific prescription of particular alternative instrumental materials and processes they can use to overcome their learning difficulties". I realized that students might profit not only from a diagnosis of their particular errors on a given test-item (see Analysis of Practice Quizzes), but also from access to some other intermediate stages of my test-development procedure. For example, the behavior-content matrix provides students with a structured summary of a chapter and forms a framework for remedial suggestions.

The list of objectives at the beginning of each study guide chapter (except Chapter 1) is just that: a transcript of the behavior-content taxonomy. The general objectives at the beginning of the list are the equivalent of the content areas of a chapter. The more specific objectives that follow are the behaviors that the student should exhibit when he/she masters the chapter. The tasks associated with each content area are listed in increasing order of cognitive levels: knowledge (K), comprehension (C), and application (A).

The Tips and Reminders section addresses topics that students find problematic or difficult as evidenced by recurrent mistakes that I encountered in my error analysis of the completion-form quizzes and in my tutoring experience.

References

Ayers, J. D. Test Item Construction. A Manual For Teachers. Edmonton: Barnett House, 1967.

Bloom, B. S. Mastery Learning, in Block, J. H. (Ed.), Mastery Learning: Theory and Practice. New York: Holt, Rinehart and Winston, 1971.

Bloom, B. S., Engelhart, M. D., Hill, H., First, E. J., & Krathwohl, D. R. Taxonomy of education objectives: Handbook I: Cognitive domain. New York: David McKay, 1956.

Brown, J. S. & Burton, R. R. Diagnostic Models for Procedural Bugs in Basic Mathematical Skills, Cognitive Science, 2, 155-192, 1978.

Ebel, R. L. Essentials of Educational Measurement. Englewood Cliffs, N.J.: Prentice-Hall, 1972.

Forsyth, G. A. A Task-First Individual-Differences Approach to Designing a Statistics and Methodology Course, Teaching of Psychology, 4, 76-78, 1977.

Medcof, J. Teaching to and Testing at four different cognitive levels. Presented at the Annual Convention of the Canadian Psychological Association, June 1978.

TO THE STUDENT

"If there _is_ no meaning", as the King said
to Alice, "that saves a world of trouble,
you know, as we needn't try to _find_ any."
(Lewis Carroll)

If "a world of trouble" describes your anticipations of the statistics course that you are beginning, you are not alone. Indeed, it has become almost a custom for statistics texts and study guides to reassure their student readers that statistics _can_ be learned and may even be _fun_.

Without offering any easy alternative to your own efforts, this study guide will try to help you understand and organize the material presented in the text, _Principles of Statistics_. To see meaning and order in statistics should hold its own rewards for you; it will also help you succeed in your course and will prepare you for possible applications of statistics in your own work or research later on.

The Parts of Each Study Guide Chapter

How can the study guide help you? To each chapter of the text there is a corresponding study guide chapter. Each study guide chapter consists of three parts:

(a) a list of _Objectives_, which provides you with a comprehensive list of tasks that you should be able to do after completing the chapter. Your instructor has received a large number of quiz questions, based on these objectives, and the exams you write may, therefore, be based on these objectives. If you are interested in the meaning of the K-C-A code of the objectives, read the preceding section "To the Instructor".

(b) a list of Tips and Reminders, each of which may (i) discuss a certain topic in greater detail than the text does, (ii) point out common sources of error (so you can avoid them), (iii) organize text material visually in a figure or chart, or (iv) familiarize you with notation, symbols, formulas, or other routine operations.

(c) a Practice Quiz, consisting of ten multiple-choice questions. These questions are similar to the quiz questions which have been given to your instructor and will increase your quiz-taking experience and confidence. The correct alternatives to each of the ten practice quiz questions are listed together with an analysis of likely errors underlying all incorrect alternatives. (Note: the review chapters A, B, C, and D do not have practice quizzes; the practice quiz in Chapter 1 is not in multiple-choice format and does not have an analysis of errors.)

How To Study Each Chapter

You should read each text chapter and the parts of the corresponding study guide chapter in the order given below. The list assumes that your instructor also assigns you a set of exercises, based on the text chapter.

1. Read the text chapter.
2. Read the study guide Objectives and Tips and Reminders.

3. Do the exercises which your instructor has assigned.
4. Take the study guide Practice Quiz.
5. If you got all the questions correct, congratulations--you can assume that you have mastered the chapter. Otherwise study the analysis which is given in the study guide for each question you got wrong. The analysis will then lead you back to certain objectives and text sections, which you should study once more, before retaking those practice-quiz questions which you got wrong.

Writing Multiple-Choice Quizzes

The questions in each practice quiz and in the set of quizzes which your instructor has received were developed especially for your text. The following advice applies both to the practice quizzes and to the quizzes your instructor has.

Do not simply choose the alternative that "looks best" to you. Almost all incorrect alternatives are based on common student errors and actual, incorrect, student answers on completion-form quizzes (quizzes in which the answer must be written out rather than an alternative chosen). This means that most incorrect alternatives have "looked good" to somebody before.

Instead, work out the given problem to the final answer; then check whether your answer is one of the listed alternatives. If not, check your calculations or revise your reasoning. You might want to check your calculations and reasoning even if your answer does coincide with one of the listed alternatives--you may have made the error on which that incorrect alternative is based. Thus, to find that your answer is one of the listed alternatives is <u>no</u> <u>guarantee</u> that your answer is correct.

The practice quiz will give you greatest benefit if you treat it like a regular quiz. Take it when you feel adequately prepared, and note the time that it takes you to complete it. Consult the quiz analysis section only <u>after</u> completing the <u>entire</u> quiz. Try to understand where you went wrong. Review all relevant material. Then retake the practice quiz questions that you answered incorrectly.

CHAPTER 1. MATH PREPARATION

OBJECTIVES

This Study Guide chapter does not directly correspond to Text Chapter 1. Instead, it is a review of all the mathematical operations and procedures that you will need in this statistics course. It should serve one of two purposes: Assure you that you are adequately prepared (high score on the practice quiz), or, guide and help you in reviewing those topics in which you find yourself weak.

In addition, some of the later Tips provide useful information on more general topics that students find troublesome, such as, for example, graphing or solving work problems.

To be adequately prepared for this course, you should be familiar with

(1) the meaning of the following operators: factorial, exponent, square root, inequality signs, absolute value, multiplication signs;

(2) the order of executing operations;

(3) operations involving exponents;

(4) operations involving fractions;

(5) operations involving factorials;

(6) the rules of rounding;

(7) decimals;

(8) percentages;

(9) signed numbers;

(10) solving equations with one unknown;

(11) solving word problems;

(12) estimating the correct answer;

(13) drawing a graph;

(14) common equivalents.

TIPS AND REMINDERS

The following tips correspond to the objectives listed above, i.e., Tip (1) is an elaboration of the skill which Objective (1) requires you to have.

(1) Operators

Factorial

The symbol $n!$ is a shorthand method of writing "multiply all numbers from this number down to the number 1".

Example

6! is read "six factorial";

6! means $6 \times 5 \times 4 \times 3 \times 2 \times 1 = 720$

Exponents

In the symbol x^n, the small n refers to the number of times x is multiplied by itself. (Note: x can represent any number.)

Example

x^4 is read "x to the fourth power";

x^4 means multiply x by itself four times, i.e., $x \cdot x \cdot x \cdot x$

Square Root

The symbol $\sqrt{x}$ represents a number that, when multiplied by itself (i.e., squared), gives you x (the number under the square root sign).

Example

$\sqrt{16}$ is read "square root of 16";

$\sqrt{16} = 4$ because $4^2 = 4 \times 4 = 16$

Inequalities

$>$ means "is greater than". For example, $6 > 2$.

$<$ means "is less than". For example, $3 < 10$.

$\geqq$ means "is greater than or equal to". For example, if $a \geqq 4$, then a can be 4, 5, 6, etc.

$\leqq$ means "is less than or equal to". For example, if $b \leqq 3$, then b can be 3, 2, 1, etc.

$a < x < b$ is read "x is greater than a and less than b"; or "x lies between a and b". For example, if $3 < x < 5$, x can be any value between 3 and 5.

$x < a$ or $x > b$ is read "x is less than a or x is greater than b". For example, if $x < 3$ or $x > 5$, x can be either less than 3 (e.g. 2, 1, 0) or x can be greater than 5 (e.g. 6, 7, 8 etc.)

Absolute Value

$|a|$ means the absolute value of a. The absolute value of any number is just the number itself, without its sign.

Examples

$|7|$ is read "the absolute value of 7" and has value 7

$|-22| = 22$

Multiplication Signs

$\times$, $\cdot$ These are multiplication symbols. Another way to indicate multiplication is to have nothing between the quantities to be multiplied. For example: $2 \times 12 = 24$; $30 \cdot 4 = 120$, and $(4)(8) = 32$.

(2) Order of Operations 

First, execute brackets (inner before outer brackets);

second, do multiplications and divisions from left to right;

third, do additions and subtractions from left to right.

Example

$3 \times 7 + \{6 - 4(8 + 5)\} = 3 \times 7 + \{6 - 4(13)\}$

$= 3 \times 7 + \{6 - 52\} = 21 + \{-46\} = -25$

(3) Operations involving Exponents

Notation

a^n : in this exponential quantity, "a" is the base and "n" is the exponent.

Example

15^3 : the base is 15; the exponent is 3

Addition/Subtraction

Exponential quantities can be added or subtracted only after the quantities have been worked out separately.

Example 1

$3^2 + 4^4 = 9 + 256 = 265$

Example 2

$4^2 + 4^2 = 16 + 16 = 32$

$4^4 = 256$ Note: $4^2 + 4^2$ is not equal to 4^4

Multiplication

(i) Squaring is multiplying a number by itself.

Example

$(7)(7) = 7^2 = 49$

(ii) Cubing is multiplying a number by itself three times.

Example

$6 \cdot 6 \cdot 6 = 6^3 = 216$

(iii) If no exponent is written above the base, the exponent is assumed to be 1.

Examples

$4 = 4^1$

$12 = 12^1$

(iv) When multiplying exponential quantities that have different bases, the quantities are worked out separately and then multiplied.

Example

$4^5 \cdot 6^2 = (1024)(36) = 36864$

(v) When exponential quantities that have the same bases are multiplied, the exponents are added to find how many times the number should be multiplied by itself.

Example

$3^5 \cdot 3 = 3^5 \cdot 3^1 = 3^{5+1} = 3^6 = 729$

(vi) Any number (other than zero) raised to the zeroth power is 1. (If both the base and the exponent are zero, the quantity is undefined.)

Examples

$7^0 = 1$

$27654^0 = 1$

Division

(i) When dividing two exponential numbers that have the same bases, take the exponent of the quantity in the numerator and subtract from it the exponent of the quantity in the denominator.

Example

$$\frac{3^{15}}{3^6} = 3^{15} \div 3^6 = 3^{(15 - 6)} = 3^9$$

(ii) In the case where the bases are different, work out the exponential quantities separately and then divide.

Example

$$\frac{4^2}{3^2} = \frac{16}{9}$$

(4) Operations involving Fractions

Notation

$\frac{a}{b}$: in this fraction, "a" is the numerator, and "b" the denominator.

Example

$\frac{2}{5}$: the numerator is 2; the denominator is 5

Addition/Subtraction

(i) When adding/subtracting fractions that have the same denominators, simply add/subtract the numerators.

Example

$$\frac{4}{7} - \frac{1}{7} = \frac{4 - 1}{7} = \frac{3}{7}$$

(ii) In the case where the denominators are different, a common denominator must be found before the operation can be carried out. The easiest way to do this is to multiply both the numerator and denominator of the first fraction by the denominator of the second fraction. Do this also for the second fraction--multiply its numerator and denominator by the denominator of the first fraction.

Example

$$\frac{5}{7} + \frac{2}{5} = \left(\frac{5}{7}\right)\left(\frac{5}{5}\right) + \left(\frac{2}{5}\right)\left(\frac{7}{7}\right) = \frac{25}{35} + \frac{14}{35} = \frac{39}{35}$$

(iii) Always remember to reduce the fraction to its lowest form. This is done by dividing both the numerator and the denominator by the same number. If unsure about which number to use, try 2, 3 or 5.

Example

$\frac{10}{20}$ To reduce the fraction to its lowest form divide both numerator and denominator by 10. $\frac{10}{20} = \frac{1}{2}$

(iv) When adding/subtracting whole numbers and fractions, it may be easiest to convert the whole number to a fraction.

Example

$$4 - 2\frac{7}{8} = \frac{32}{8} - \frac{23}{8} = \frac{9}{8}$$

Multiplication

(i) When multiplying fractions, multiply the numerators together and then multiply the denominators.

Example

$$\left(\frac{4}{5}\right)\cdot\left(\frac{3}{6}\right) = \frac{4 \times 3}{5 \times 6} = \frac{12}{30}$$ This fraction can be reduced by dividing both the numerator and denominator by 6.

$$\frac{12}{30} = \frac{2}{5}$$

(ii) When multiplying a whole number by a fraction, multiply the numerator by the whole number and divide by the denominator. (See

Example 1.) Another way to do this is to follow the method in (i) by first changing the whole number to a fraction using 1 as the denominator (See Example 2.)

Example 1

$$6 \cdot \frac{2}{7} = \frac{6 \times 2}{7} = \frac{12}{7}$$

Example 2

$$6 \cdot \frac{2}{7} = \left(\frac{6}{1}\right)\left(\frac{2}{7}\right) = \frac{6 \times 2}{1 \times 7} = \frac{12}{7}$$

Division

(i) When dividing two fractions, take the second fraction and invert it (put the denominator on top and numerator on the bottom). Then multiply the two fractions. (Note: If a number in the numerator is also shown in the denominator, these numbers can be cancelled, leaving 1 in their place. See Example 1.)

Example 1

$$\frac{2}{5} \div \frac{4}{5} = \frac{2}{\cancel{5}_1} \times \frac{\cancel{5}^1}{4} = \frac{2 \times 1}{1 \times 4} = \frac{\cancel{2}^1}{\cancel{4}_2} = \frac{1}{2}$$

Example 2

$$\frac{3}{7} \div 7 = \frac{3}{7} \times \frac{1}{7} = \frac{3 \times 1}{7 \times 7} = \frac{3}{49}$$

(ii) Cancel only if there is a multiplication or division sign. DO NOT CANCEL ACROSS AN ADDITION OR SUBTRACTION SIGN. In this case, you should cancel only after the addition or subtraction has been completed.

Example 1

$$\left(\frac{4}{5}\right)\left(\frac{5}{3}\right) = \frac{4}{\not{5}} \times \frac{\not{5}}{3} = \frac{4}{3}$$

Here it is correct to cancel the 5s.

Example 2

$$\frac{4 + 3}{2}$$

Here it is incorrect to cancel the 2 and take 2 out of the 4.

Incorrect: $\frac{\not{4}^{2} + 3}{\not{2}_{1}} = 5$ Correct: $\frac{4 + 3}{2} = \frac{7}{2}$

Example 3

Note: Don't forget that the line between the numerator and the denominator means "divide".

$$\frac{\overbrace{\frac{3}{5} + \frac{1}{5}}^{\text{numerator}}}{\underbrace{\frac{6}{3} + \frac{1}{3}}_{\text{denominator}}} = \frac{\frac{4}{5}}{\frac{7}{3}} = \frac{4}{5} \div \frac{7}{3} = \frac{4}{5} \times \frac{3}{7} = \frac{12}{35}$$

(5) Operations involving Factorials

Addition/Subtraction

Factorials can only be added or subtracted after they have been worked out separately.

Examples

$3! + 4!$ is not equal to $7!$

$3! + 4! = 3 \cdot 2 \cdot 1 + 4 \cdot 3 \cdot 2 \cdot 1 = 6 + 24 = 30$

$7! = 7 \cdot 6 \cdot 5 \cdot 4 \cdot 3 \cdot 2 \cdot 1 = 5040$

Multiplication

First work out the factorials, then multiply the quantities together.

Example

$5! \times 4! = (5 \cdot 4 \cdot 3 \cdot 2 \cdot 1)(4 \cdot 3 \cdot 2 \cdot 1) = (120)(24) = 2880$

Division

When dividing factorials, write them out in long form, cancel any identical terms, and then complete the calculation.

Example 1

$$\frac{10!}{8!} = \frac{10 \cdot 9 \cdot \cancel{8} \cdot \cancel{7} \cdot \cancel{6} \cdot \cancel{5} \cdot \cancel{4} \cdot \cancel{3} \cdot \cancel{2} \cdot \cancel{1}}{\cancel{8} \cdot \cancel{7} \cdot \cancel{6} \cdot \cancel{5} \cdot \cancel{4} \cdot \cancel{3} \cdot \cancel{2} \cdot \cancel{1}} = \frac{10 \times 9}{1} = 10 \times 9 = 90$$

Example 2

$$\frac{8!}{10!} = \frac{\cancel{8} \cdot \cancel{7} \cdot \cancel{6} \cdot \cancel{5} \cdot \cancel{4} \cdot \cancel{3} \cdot \cancel{2} \cdot \cancel{1}}{10 \cdot 9 \cdot \cancel{8} \cdot \cancel{7} \cdot \cancel{6} \cdot \cancel{5} \cdot \cancel{4} \cdot \cancel{3} \cdot \cancel{2} \cdot \cancel{1}} = \frac{1}{10 \times 9} = \frac{1}{90}$$

Special Value

Note that $0! = 1$.

Example

$$\frac{4!}{0!} = \frac{4 \cdot 3 \cdot 2 \cdot 1}{1} = 24$$

(6) Rounding

When rounding a number, consider how far it is to be rounded (for example, to 2 decimal places). Look at the next number (for example, the third decimal place). If this number is 6, 7, 8, or 9 you round up. If this number is 0, 1, 2, 3, or 4, you round down. If this number is 5 followed by anything but 000 . . . , you round up. If the digits following the 5 are exactly 000 . . . , or there are no digits following the 5, you round the last retained digit to an even number. See Text Table 3.3 and Study Guide Chapter 3, Tip A for further examples of these rounding rules.

Example 1

Round 3.78643 to 3 decimal places.

Solution: 3.786|43 The first digit to be dropped is 4; therefore, rounding down to 3 decimal places gives 3.786.

Example 2

Round 253.836 to the nearest ten.

Solution: 25|3.836 The first digit to be dropped is 3; therefore, rounding down to the nearest ten gives 250.

Example 3

Round 94976.3 to the nearest hundred.

Solution: 949|76.3 The first digit to be dropped is 7; therefore, rounding up to the nearest hundred gives 95000.

Example 4

Round 32.55 to one decimal place.

Solution: 32.5|5 — The first digit to be dropped is 5 and is not followed by zeros; therefore, rounding up to one decimal place gives 32.6.

Example 5

Round 0.365 to two decimal places.

Solution: 0.36|5 — The first digit to be dropped is 5 and no digits follow it, so the last digit retained must be an even number. Therefore, 0.365 rounded to two decimal places gives 0.36.

(7) Decimals

Changing Fractions to Decimals

To change a fraction to a decimal, divide the denominator into the numerator.

Examples

$$\frac{2}{10} = 2 \div 10 = 0.2$$

$$\frac{3}{8} = 3 \div 8 = 0.375$$

(8) Percentages

Notation

% : this symbol means "per hundred".

Example

10% : means "10 per hundred" and is read as "10 percent".

Changing Decimals to Percentages

To change a decimal into a percentage, multiply the decimal by 100.

Example 1

Change 0.437 into a percentage.

Solution: $0.437 \times 100 = 43.7\%$

Example 2

Change $\frac{2}{16}$ into a percentage.

Solution: a) Change the fraction to a decimal: $\frac{2}{16} = 0.125$

b) Multiply the decimal by 100: $0.125 \times 100 = 12.5\%$

Changing Percentages to Decimals

To change a percentage to a decimal divide the percentage by 100.

Example

$3\% = \frac{3}{100} = 0.03$

Application

In some problems you may be asked to find a certain percentage of a number or quantity. A way to do this, is to change the percentage to a decimal and then to multiply the number by this decimal.

Example

What is 35% of $600?

Solution: a) $35\% = \frac{35}{100} = 0.35$

b) $0.35 \times 600 = 210$

c) Therefore 35% of $600 is $210

(9) Signed Numbers

Multiplication

When multiplying positive/negative numbers, the following rules apply:

$+ \times + = +$	$7 \times 8 = 56$
$- \times - = +$	$(-10)(-5) = 50$
$+ \times - = -$	$7 \times (-6) = -42$
$- \times + = -$	$(-7) \times 6 = -42$

Division

When dividing positive/negative numbers, the following rules apply:

$+ \div + = +$	$12 \div 2 = 6$
$- \div - = +$	$-10 \div (-5) = 2$
$+ \div - = -$	$8 \div (-4) = -2$
$- \div + = -$	$-8 \div 4 = -2$

The rules for multiplication and division can be summarized as follows:

same signs = positive number

different signs = negative number

(10) Equations with One Unknown

The most important point in working with equations is to understand that the unknown quantity (frequently called x) must be isolated on one side of the equation and that all the other numbers must be placed on the other side of the equation. Study the following examples to see more clearly how a solution is found. Note: Any operation done to one side of the equation (e.g., subtract 12), must be also done to the other side of the equation. Remember to check your final answer by substitution, i.e., put your final answer into the original equation and work through.

Example 1

$x + 12 = 7$

Solution: Subtract 12 from both sides of the equation.

$$x + 12 - 12 = 7 - 12$$

$$x = -5$$

Check: $-5 + 12 = 7$

$$7 = 7$$

Example 2

$$\frac{x}{4} + 3 = 23$$

Solution: a) Subtract 3 from both sides of the equation.

$$\frac{x}{4} + 3 - 3 = 23 - 3$$

$$\frac{x}{4} = 20$$

b) Multiply both sides by 4.

$$\left(\frac{x}{4}\right)(4) = (20)(4)$$

$$x = 80$$

Check: $\frac{80}{4} + 3 = 23$

$20 + 3 = 23$

$23 = 23$

Example 3

$$\frac{7x}{3} - 14 = 37$$

Solution: a) Add 14 to both sides of the equation.

$$\frac{7x}{3} = 37 + 14 = 51$$

b) Multiply both sides of the equation by $\frac{3}{7}$

$$\left(\frac{3}{7}\right)\left(\frac{7x}{3}\right) = \left(\frac{3}{7}\right)(51)$$

$$x = \frac{3 \times 51}{7} = \frac{153}{7}$$

Check:

$$\frac{7\left(\frac{153}{7}\right)}{3} - 14 = 37$$

$$\frac{153}{3} - 14 = 37$$

$$51 - 14 = 37$$

$$37 = 37$$

Example 4

$$12 = \sqrt{\frac{(144)(2)}{x}}$$

Solution: a) Multiply values under the square root sign.

$$12 = \sqrt{\frac{288}{x}}$$

b) Remove the square root signs by squaring both sides.

$$(12)^2 = \frac{288}{x}$$

$$144 = \frac{288}{x}$$

c) Multiply both sides of equation by x (i.e., cross-multiply).

$$144x = 288$$

d) Isolate x on left side by dividing both sides by 144.

$$\frac{144x}{144} = \frac{288}{144}$$

$$x = \frac{288}{144} = 2$$

Check:

$$12 = \sqrt{\frac{(144)(2)}{2}}$$

$$12 = \sqrt{\frac{288}{2}}$$

$$12 = \sqrt{144}$$

$$12 = 12$$

<u>Example 5</u>

$$9 = \frac{4}{\sqrt{(32)(3)}} \sqrt{x}$$

Solution: a) Multiply values under the square root sign.

$$9 = \frac{4}{\sqrt{96}} \sqrt{x}$$

b) Remove the square root signs by squaring both sides.

$$(9)^2 = \frac{(4)^2}{96} x$$

$$81 = \frac{16x}{96}$$

c) Multiply both sides of equation by 96 (i.e., cross-multiply).

$$16x = 81 \times 96$$

$$16x = 7776$$

d) Isolate x on left side by dividing both sides by 16.

$$\frac{\cancel{16}x}{\cancel{16}} = \frac{7776}{16}$$

$$x = \frac{7776}{16} = 486$$

Check:

$$9 = \frac{4}{\sqrt{(32)(3)}} \sqrt{486}$$

$$9 = \frac{4}{\sqrt{96}} \sqrt{486}$$

$$(9)^2 = \frac{4^2}{96} (486)$$

$$81 = 81$$

(11) Solving Word Problems

Every word problem is different. The only way you can master them is through practice. The best way to cope with a word problem is to forget about all the words and rewrite the problem in terms of numbers and equations. In this way you extract all the important information and disregard all the details that only add to the confusion. The following technique for finding a solution is suggested:

(a) DRAW A DIAGRAM. A diagram works like a map. In the diagram you note all the given information and what you are asked to find. It describes the entire word problem but uses only the necessary words; therefore, it is easier to understand.

(b) LIST ANY INFORMATION THAT DEALS WITH NUMBERS OR QUANTITIES. Check that the numbers you have written are the same ones used in the problem. This will avoid any careless errors. From this step you can start to set up the equations.

(c) SET UP AND SOLVE THE EQUATIONS. Be sure to isolate the unknown quantity on one side of the equation. [For further information and examples on this, see Tip (10)].

(d) CHECK YOUR ANSWER. This will allow you to catch any errors and correct them.

(e) WRITE A CONCLUDING STATEMENT. Remember, the word problem is given to you in words. Therefore, your final answer should also be given in words, stating exactly what it was that you were supposed to find.

Illustration of Technique

Example 1

The perimeter (distance measured around the outer edge) of a rectangle is 208 cm. The length is 18 cm longer than the width. What are the dimensions of the rectangle?

Solution: a) Diagram

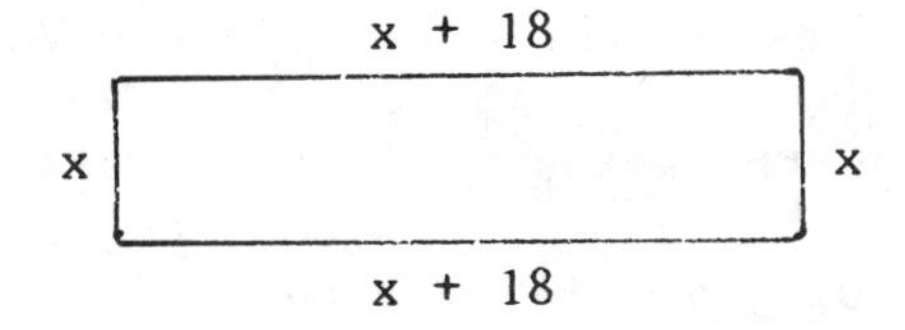

b) <u>List Information</u>

perimeter = 208 cm

width = x

length = x + 18

Therefore, $x + (x + 18) + x + (x + 18)$

$= 2x + 2(x + 18) = 208$

c) <u>Solve Equation</u>

$x + x + 18 + x + x + 18 = 208$

$4x = 208 - 36$

$$x = \frac{172}{4} = 43$$

Therefore the width = 43 cm and the length = 43 + 18

= 61 cm

d) <u>Check Answer</u>

$43 + 61 + 43 + 61 = 208$

Therefore the answer must be correct.

e) <u>Concluding Statement</u>

The width of the rectangle is 43 cm and its length is 61 cm.

<u>Example 2</u>

A student's scores on three tests are 91%, 65%, and 88%. What must she score on her fourth test so that her average will be 80%?

Solution: a) <u>Diagram</u> (in this case, it is a chart)

Score on test				Average
91	65	88	x	80

b) List Information

$$\text{average} = \frac{\text{sum of scores}}{\text{number of items}}$$

$$80 = \frac{91 + 65 + 88 + x}{4}$$

c) Solve Equation

$$80 = \frac{244 + x}{4}$$

$$4 \times 80 = \frac{244 + x}{4} \times \frac{4}{1}$$

$$320 = 244 + x$$

$$320 - 244 = x$$

$$x = 76$$

d) Check Answer

$91 + 65 + 88 + 76 = 320$

$320 \div 4 = 80$

Therefore the answer must be correct.

e) Concluding Statement

The student must receive a score of 76% on her fourth test in order to get an average of 80%.

Example 3

A baseball diamond is a square tipped on one of its corners. The length of one of its sides is 25 meters. What is the distance, in a straight line, from second base to home? (Hint: The hypotenuse of a right-angled triangle is "c" in the formula $c^2 = a^2 + b^2$, where a and b are the lengths of the other two sides.)

Solution: a) Diagram

See Figure SG 1.1.

Figure SG 1.1 Baseball Diamond

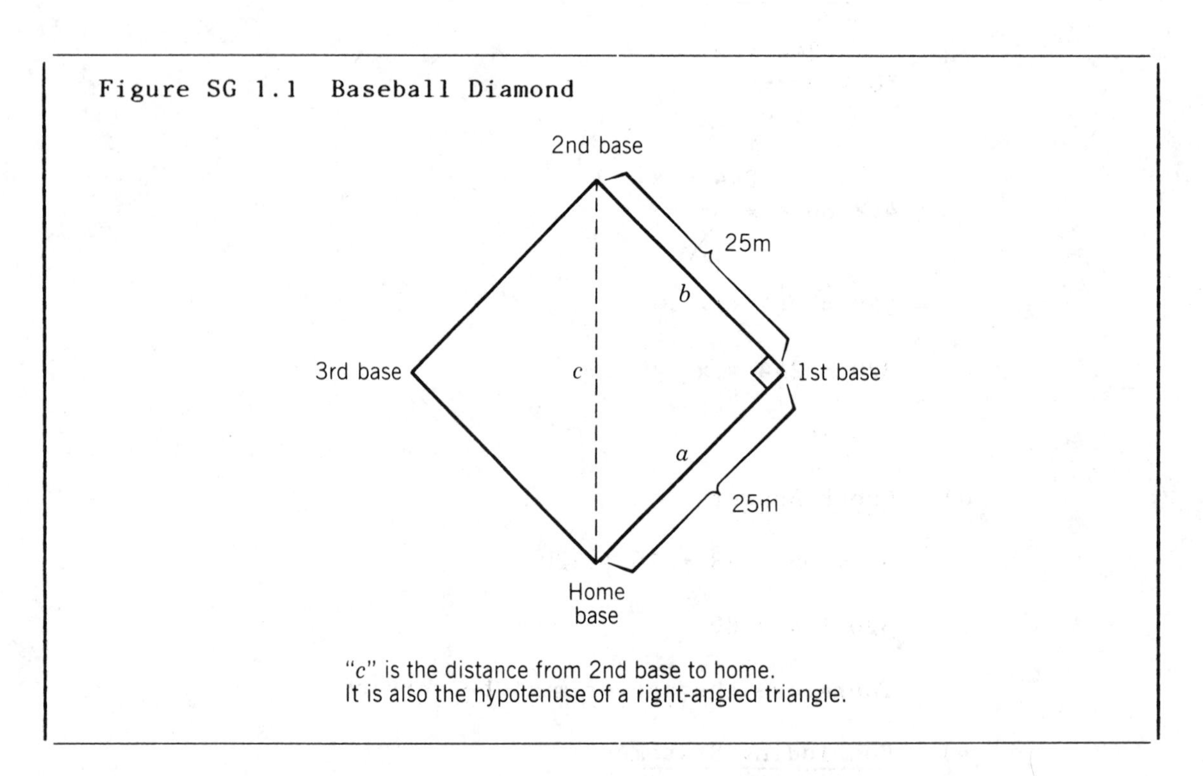

"c" is the distance from 2nd base to home.
It is also the hypotenuse of a right-angled triangle.

b) List Information

The length of each side is 25 meters.

The length of c is wanted.

$c^2 = a^2 + b^2$ where $a = 25$ and $b = 25$

Therefore $c^2 = (25)^2 + (25)^2$

c) Solve Equation

$$c^2 = (25)^2 + (25)^2$$
$$= 625 + 625$$
$$= 1250$$

Therefore, $c = \sqrt{1250} = 35.355$ or 35 meters

d) Check Answer

$$c^2 = a^2 + b^2$$

left side: $c^2 = (35.355)^2$

$= 1249.976$ or 1250

right side: $a^2 + b^2$

$= (25)^2 + (25)^2$

$= 625 + 625$

$= 1250$

left side = right side

Therefore the answer must be correct.

e) Concluding Statement

This distance from second base to home base in a straight line is approximately 35 meters.

(12) Estimation

Estimating the answer before you actually carry out the calculation is a useful tool in determining whether or not the calculation was done properly.

If your estimated and calculated answers are totally different, it indicates that something was not done correctly and you should attempt the problem again. Remember that an estimate is just that--a rough approximation of the answer. It is not meant to take the place of the correct answer.

Estimating Percentages

In some problems you may be asked to find a certain percentage of a number. One method of estimating the answer is to use fractions such as 1/2, 1/4, 1/10, 1/100, in order to find boundaries of an interval in which the answer must fall.

Example 1

What is 17% of 48?

Estimate: 17% lies between 10% (which is 1/10) and 25% (which is 1/4).

$$\left(\frac{1}{10}\right)(48) = 4.8 \qquad \left(\frac{1}{4}\right)(48) = 12$$

Therefore the answer will be between 4.8 and 12.

Exact Answer:

$$17\% \text{ of } 48 = \frac{17}{100} \times 48 = .17 \times 48 = 8.16$$

Example 2

Find 43% of 60.

Estimate: 43% lies between 25% (which is 1/4) and 50% (which is 1/2).

(1/4)(60) = 15

(1/2)(60) = 30

Therefore the answer will be between 15 and 30.

Exact Answer:

$$43\% \text{ of } 60 = \frac{43}{100} \times 60 = .43 \times 60 = 25.8$$

Estimating Other Answers

In other calculations you may not be using percentages. In this case, round each number to one significant digit followed by zeros and then perform the operation.

Example 1

Compute: $24 \times .037$

Estimate: round 24 to 20

round .037 to .04

$20 \times .04 = .80$

Exact Answer: $24 \times .037 = .888$

Example 2

Compute: $38 \div 12.2$

Estimate: round 38 to 40

round 12.2 to 10

$40 \div 10 = 4$

Exact Answer: $38 \div 12.2 = 3.115$

(13) Drawing a Graph

(i) Your figure should not be so small that the numbers cannot be read nor so large that it takes up an entire page.

(ii) Use a ruler--it helps avoiding careless mistakes.

(iii) When adding a scale, keep the distance between numbers constant. Don't use 1 inch to separate the numbers 1 and 2, and then use half an inch to separate the numbers 2 and 3.

(iv) The axes of a graph are simply two numberlines placed perpendicular to each other.

(v) Be sure that your graph answers all of the following questions: What is being measured on the abscissa (or x-axis)? What is being measured on the ordinate (or y-axis)? What are the scales? What is the title of the graph? See Figures SG 1.2 and SG 1.3.

Figure SG 1.2 Essentials of a Graph

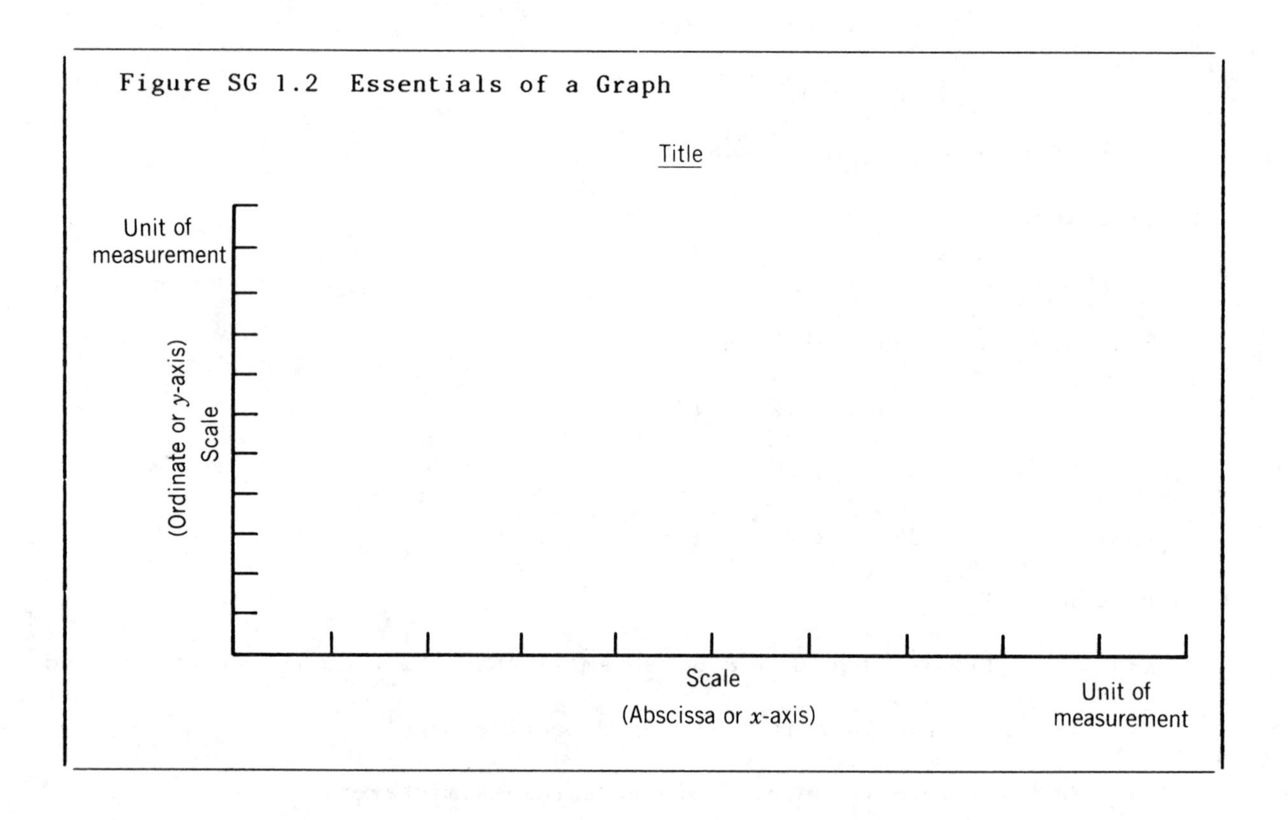

Figure SG 1.3 Example of a Graph

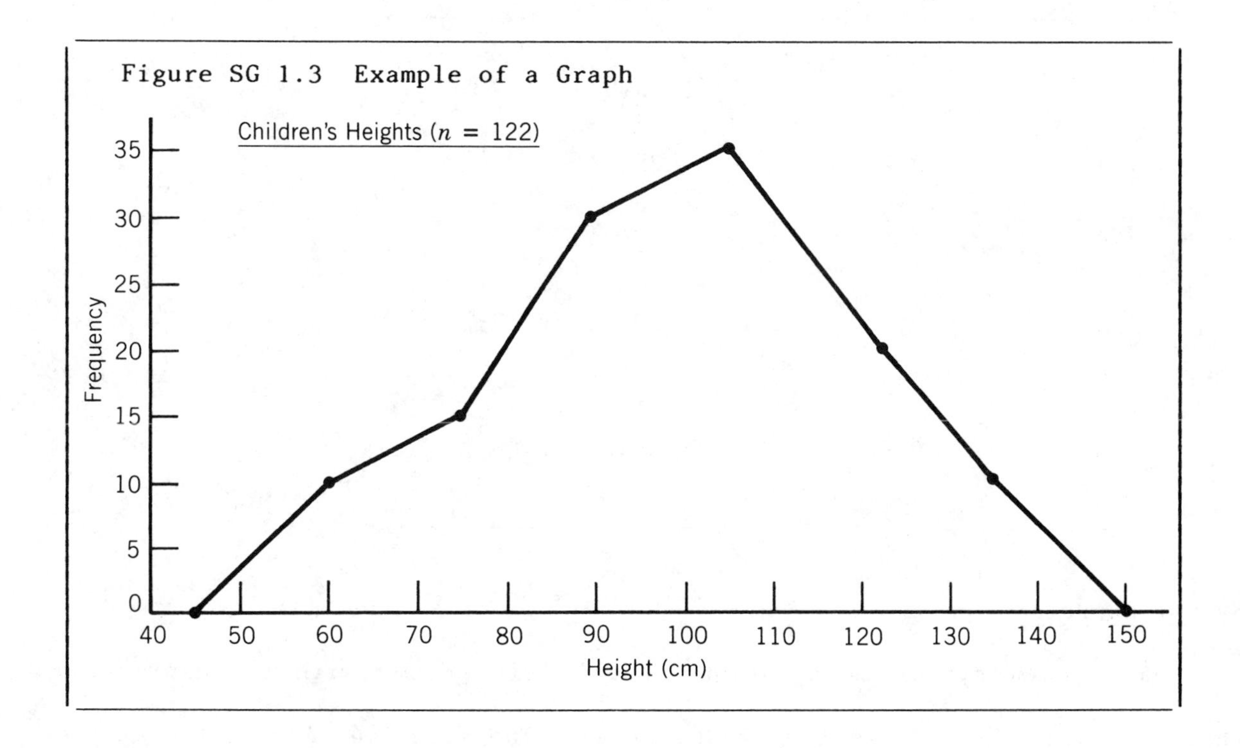

(14) Table of Common Equivalents

The quantities in each line of the table are equivalents and are often used interchangeably. It would save you time to be quite familiar with the table.

FRACTION	DECIMAL	PERCENTAGE
1/2	0.5	50%
1/4	0.25	25%
3/4	0.75	75%
1/3	0.333	33.3% or $33\frac{1}{3}\%$
2/3	0.666	66.6% or $66\frac{2}{3}\%$

Symbol Exercise

Match each symbol or equation in the left column with the expression that best describes it in the right column. The solutions are listed at the end of the chapter.

___ 1.	$\|-14\|$	a.	x cubed
___ 2.	$x > 7$	b.	x lies between 5 and 10
___ 3.	x^3	c.	twelve factorial
___ 4.	$x \leqq 1$	d.	x is greater than or equal to one
___ 5.	$12!$	e.	x is greater than 7
___ 6.	$5 < x < 10$	f.	x is less than 7
___ 7.	$\sqrt{12}$	g.	x is less than 5 or x is greater than 10
___ 8.	$x \geqq 1$	h.	twenty-one times five
___ 9.	$(14)(3)$	i.	x is less than or equal to one.
___ 10.	$x < 5$ or $x > 10$	j.	the absolute value of negative fourteen

f 11. $x < 7$ k. the square root of 12

h 12. $21 \cdot 5$ ℓ. fourteen times three

PRACTICE QUIZ

1. Round 27.36793 to 2 decimal places. 27.37

2. Change $\frac{4}{16}$ to a percentage. .25

3. $3^3 \cdot 4^3 + 7^8 \div 7^5 = ?$

   $27 \times 64 + 343 = 2071$

4. Change 23% to a decimal. 0.23

5. State whether the answer will be positive or negative. Do not calculate the answer:

   a) $\frac{-325}{4}$ − b) $(-15)(-23)$ +

6. Estimate the answers to the following questions. Do not calculate the exact answers.

   a) $23 \times .0573 = ?$ 1.5 ≈1.5

   b) Find 66% of 80. b/w 40 - 60

7. $\left(\frac{12}{7} + \frac{4}{3}\right) \times 1\frac{7}{8} = ?$ $\left(\frac{36+28}{21}\right) \times \frac{15}{8} = \frac{64}{21} \times \frac{15}{8} \rightarrow \frac{960}{168} \rightarrow \frac{120}{21} = \frac{40}{7}$ $5\frac{5}{7}$

8. $4(23 + 7) + 3 \times 5 - 8(10 + 4) = ?$ $= 120 + 15 - 112 = 23$

9. Round 1664.3998 to the nearest hundred. 1700

10. $3! \times 6! \div 5! = ?$ $(3\times2)(6\times5\times4\times3\times2)\div(5\times4\times3\times2)$ $6 \times 720 = 4320/120 = 36$

11. Solve for x:

    $15 = 1.02\sqrt{x}$

    $\frac{15}{1.02} = 14.71 = \sqrt{x}$ ± 216.26

12. $(21 \div 3)\ (3 \times 7) + 8 \times 2 - 6 = ?$

13. Change $\frac{15}{20}$ to a decimal

14. $(5^2 + 3^2 \div 3^0) \div 4^4 = ?$

15. $$\frac{\left(\frac{10 + 8}{5}\right) \div \left(\frac{1}{3}\right)}{\frac{7}{3} \quad \frac{3}{5}} = ?$$

16. $5! + 6! - 4! = ?$

17. What is 44% of $1200?

18. The perimeter of a square is 48 cm. What is its area (i.e., base times height)? (<u>Note</u>: your answer should be given in square cm.)

<u>Symbol Exercise Solutions</u>

1j, 2e, 3a, 4i, 5c, 6b, 7k, 8d, 9ℓ, 10g, 11f, 12h.

<u>Practice Quiz Answers</u>:

1) 27.37

2) 25%

3) 2071

4) 0.23

5) a) negative

   b) positive

6) a) any number between 1.0 and 1.5

b) between 40 and 60

7) 40/7

8) 23

9) 1700

10) 36

11) 216.26

12) 157

13) 0.75

14) 17/128

15) 81/13

16) 816

17) $528

18) 144 square cm

CHAPTER 2. DISTRIBUTIONS

OBJECTIVES

Chapter 2 of the text discusses how to represent and summarize data effectively. You should be familiar with the following topics:

- I the concept of a variable;
- II frequency distributions;
- III the summarization principle;
- IV the description of a distribution by its shape;
- V technical details of the graphing of a distribution.

More specifically, after reading the text you should be able to

I K (1) define the term *variable* (Text 2.4);

C (2) distinguish between a continuous and a discrete variable (Text 2.4);

II K (3) define the terms *frequency distribution*, *grouped frequency distribution*, *relative frequency*, *interval*, *real limits*, *apparent limits* (Text 2.1, 2.2, 2.3, 2.4, 2.6);

C (4) distinguish between the real and apparent limits of an interval (Text 2.4);

(5) distinguish between frequency and relative frequency (Text 2.6, 2.7);

A (6) determine the real limits of a score or interval (Text 2.4; Study Guide Tip A);

(7) determine the width and the midpoint of a given real or apparent interval (Text 2.4; Study Guide Tips A and B);

III K (8) list the three component principles of the summarization principle (Text 2.3);

A (9) apply the principle that statistical summaries should be useful, to decide on

(a) the use of a graph or a table to represent a set of data (Text 2.2);

(b) the use of a histogram or a polygon (Text 2.2);

(c) whether or not to group a given set of data when constructing a frequency distribution (Text 2.3);

(d) the number of intervals to use for a grouped-frequency distribution (Text 2.3);

(10) apply the principle of fairness of representation to decide on when to use simple frequencies as opposed to relative frequencies to represent a set of data (Text 2.6);

(11) apply the principle of ease of representation to decide on

(a) the limits of intervals in grouped-frequency distributions (Text 2.3);

(b) whether to represent a grouped-frequency distribution by its real or apparent limits in a histogram, or, by its real or approximate interval midpoints in a polygon (Text 2.4);

(c) when to represent a given set of data as either a univariate scatterplot, a frequency distribution, or a grouped-frequency distribution (Text 2.5; Figure SG 2.1);

IV K (12) define the terms idealized distribution and empirical distribution, symmetric, nonsymmetric, unimodal, bimodal, uniform, skewed-to-the-right (left), positively (negatively) skewed (Text 2.7);

A (13) describe the shape of a distribution (Text 2.7);

V K (14) define the terms abscissa, ordinate (Text 2.2);

C (15) understand the conventional use of the two axes when graphing a frequency distribution (Text 2.2);

(16) interpret the representations of frequency in the different graphs (histogram, polygon, univariate scatterplot) (Text 2.2, 2.5, 2.6).

Figure SG 2.1 Decision Chart for Data Representation

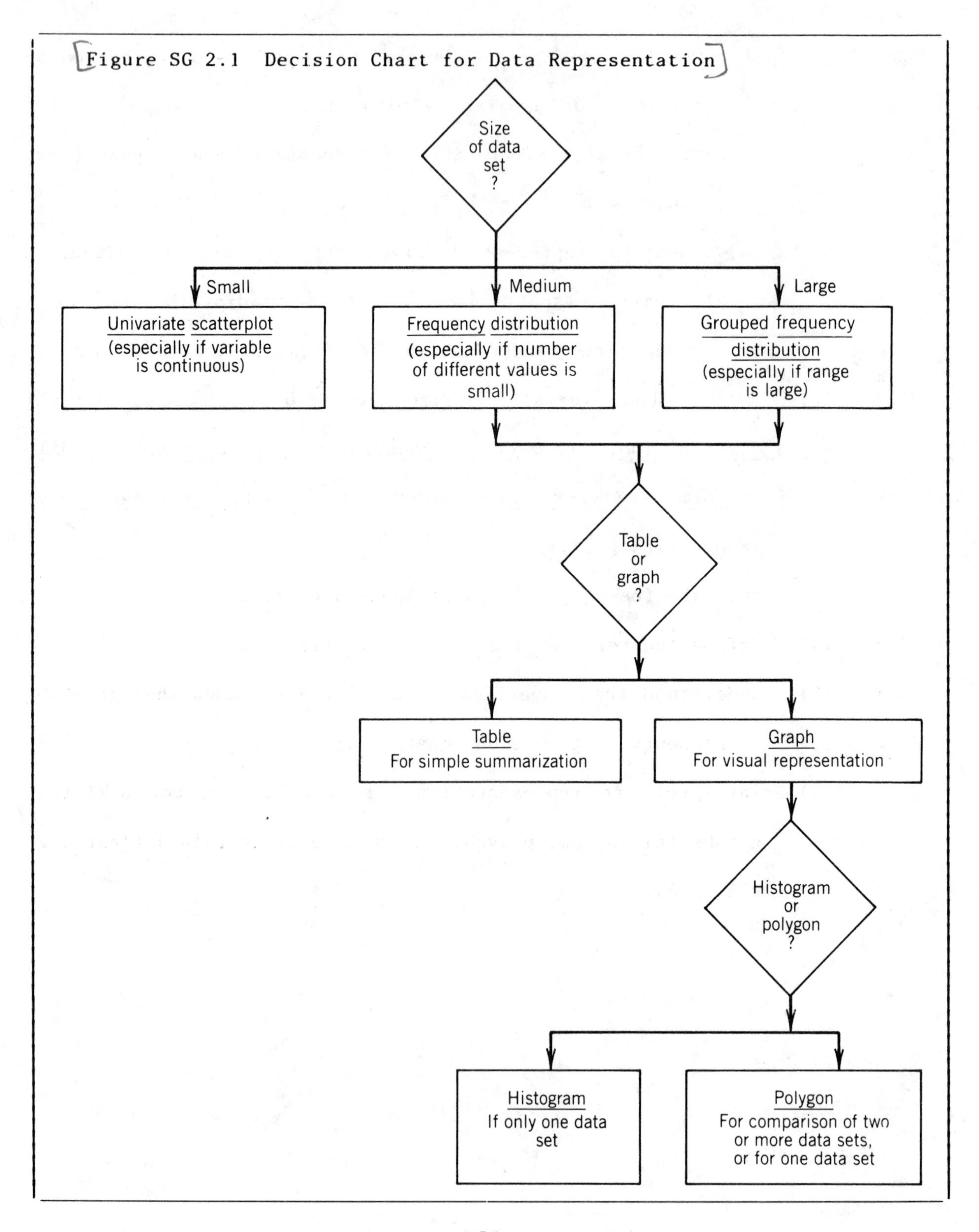

TIPS AND REMINDERS

A. Real Limits

Section 2.4 of the text introduces the notion of real limits and discusses the reasons for their use. Individual scores as well as intervals have real limits. What these limits are depends on the precision of the individual scores. The following examples will illustrate that the real limits of a score or interval are quite "natural" and easy to determine. To find the real limits of a particular score, take half of the smallest possible difference between two scores of the set and subtract it from the score you are interested in to find its lower real limit, or add it to the score to find its upper real limit. The smallest possible difference between two scores depends on the precision of the individual score, i.e., the number of decimals to which they are reported. For integer scores, the smallest possible difference is 1; scores that are reported to one decimal can differ by 0.1, and so on.

Set of scores	.. 2, 4, 5, 8, 9, ..	..3.9, 4.0, 4.3, 4.7, 5.0, ..	.. 1.20, 1.23, 1.26, 1.31, 1.35, ..
Smallest possible difference	1	0.1	0.01
Example score	5	5.0	1.23
Real limits	4.5 - 5.5	4.95 - 5.05	1.225 - 1.235

The real limits of an interval are the lower real limit of the lowest score of the apparent interval and the upper real limit of the highest score of the apparent interval. For the above examples:

Apparent interval	5 - 9	4.0 - 4.9	1.20 - 1.29
Real limits of interval = Real interval	4.5 - 9.5	3.95 - 4.95	1.195 - 1.295
Width of interval	5.0	1.00	0.100

Note that the width of an interval is defined as the width of the real interval.

B. Midpoints

The midpoint of an interval is easy to determine. The midpoint is halfway between the upper and lower limits of the interval. (These limits may be either the apparent or the real limits.) To find the halfway point between two limits, just add the numbers and divide by two. For example, the midpoint of the apparent interval 5 - 9 is $(5 + 9)/2 = 14/2 = 7$. The midpoint of the real interval 3.95 - 4.95 is $(3.95 + 4.95)/2 = 8.90/2 = 4.45$.

Since the width of an interval is defined to be the width of the real interval, the distance of the midpoint from either real limit is one-half the width. In the real interval 3.95 - 4.95, the width is 1.0; the midpoint (4.45) is half this distance, that is 0.5, from the lower real limit (3.95) and also 0.5 from the upper real limit (4.95).

C. Axes of Graphs

On any graph, always label both axes, that is, indicate the name of the variable that each axis represents and its values.

In frequency distribution graphs or relative-frequency graphs the abscissa shows the range of values of a given variable; their frequencies or relative frequencies are represented on the ordinate. While the values of the variable often do not start with or include zero, the frequency and relative frequency scales always have to originate in zero (see Figures 2.1 to 2.4 and 2.8 in the text). Graphs shown in advertisements often violate this "rule" which is an application of the principle of fairness of representation. Compare the impressions conveyed by the two graphs in Figure SG 2.2.

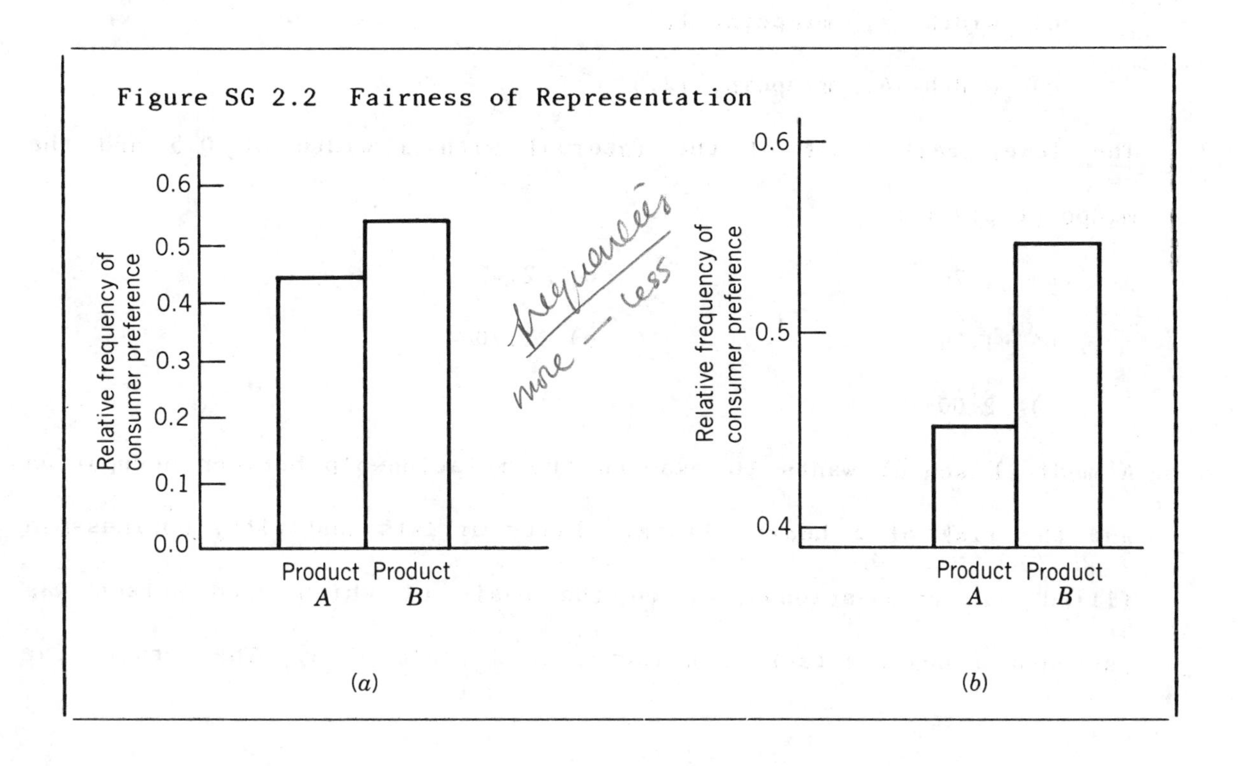

Figure SG 2.2 Fairness of Representation

PRACTICE QUIZ

Questions

1. The real limits of the score 1.0 are:

   a) 0.5 - 1.5

   b) 0.95 - 1.05

   c) 0.995 - 1.005

   d) 0.0 - 1.0

   e) 1.0 - 2.0.

2. What are the width and the midpoint of the apparent interval 10 - 14?

   a) width 5, midpoint 12.5

   b) width 5, midpoint 12

   c) width 4, midpoint 11.5

   d) width 4, midpoint 12

   e) width 4, midpoint 12.5.

3. The lower real limit of the interval with a width of 0.5 and the midpoint 2.2 is:

   a) 1.70

   b) 1.95

   c) 2.00

   d) 2.45

   e) 2.70

4. A medical school wants to examine the relationship between occupation and the risk of a heart attack. Fifty artists and fifty businessmen filled out a questionnaire, on the basis of which each person was assigned a heart attack risk factor on a scale of 5. The results for

the sample of artists are summarized below as a frequency distribution while the results for the sample of businessmen are represented by a relative frequency distribution.

Artists (N = 50)		Businessmen (N = 50)	
Risk Factor	Frequency	Risk Factor	Relative Frequency
1	6	1	.10
2	13	2	.20
3	12	3	.26
4	10	4	.26
5	9	5	.18
	50		1.00

What can you say about the frequency of high risk individuals (Factor 5) among businessmen as compared to artists?

a) Factor 5 frequency is higher among businessmen.

b) Factor 5 frequency is lower among businessmen.

c) Factor 5 frequency is the same for businessmen and artists.

d) Cannot be determined without further information.

5. Compare the relative frequency polygons, A and B, in Figure SG 2.3. The sample sizes of A and B are 40 and 70, respectively. What can you say about the frequency of score 53?

a) greater in distribution A than in distribution B.

b) smaller in distribution A than in distribution B.

c) equal in distribution A and distribution B.

d) cannot be determined without further information.

Figure SG 2.3 Practice Quiz, Question 5

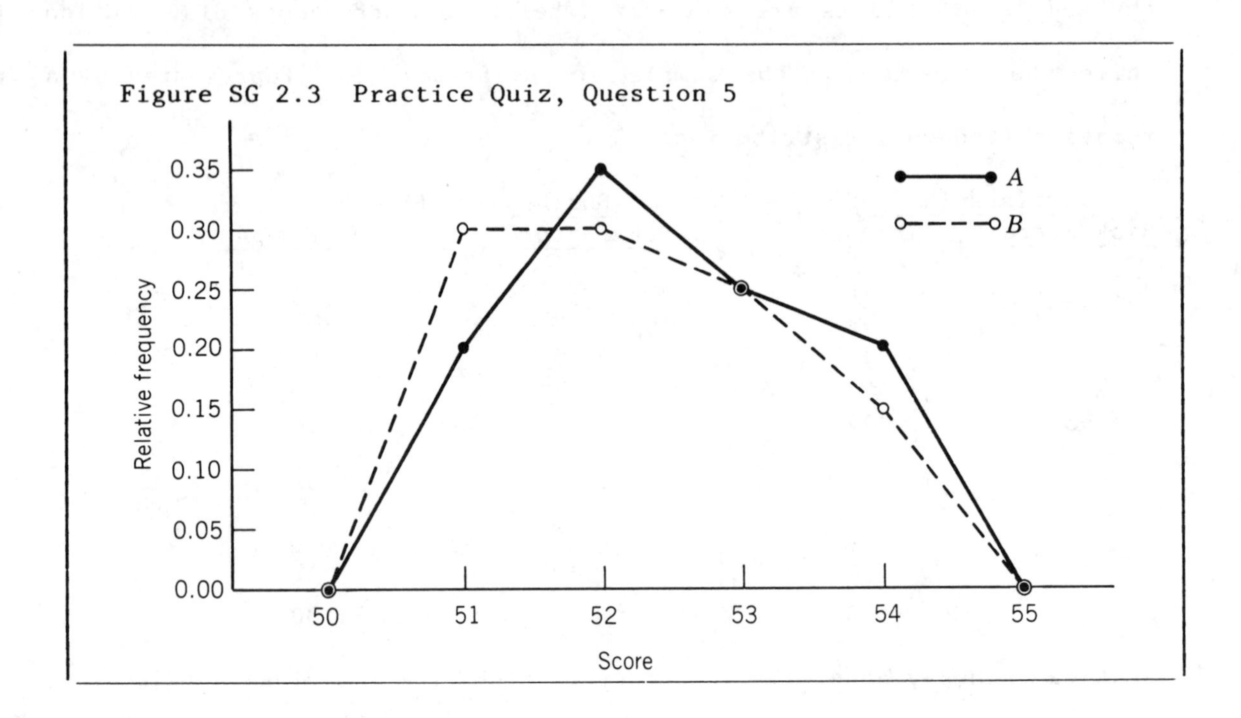

6. A university wants to compare the college entrance exam scores of students that entered in 1960 and in 1970. A grouped frequency polygon of the exam scores for each of the two entrance years is drawn. In both years, 1400 students were admitted. Should simple frequencies or relative frequencies be used in plotting the two polygons?

   a) Simple frequencies, because relative frequencies would change the shape of the distributions.

   b) Relative frequencies, because simple frequencies would change the shape of the distribution.

   c) Simple frequencies, because the two groups have equal sample sizes.

d) Relative frequencies, because the two groups have equal sample sizes.

e) Either simple or relative frequencies, because the two groups have equal sample sizes.

7. In light of the summarization principle what would be the most appropriate way to represent the following set of data?

17.18 15.97 16.04 16.53 16.61 17.02 16.25 17.11
16.78 16.99 17.00.

a) univariate scatterplot

b) frequency histogram

c) frequency polygon

d) grouped-frequency histogram

e) grouped-frequency polygon.

8. In light of the summarization principle what would be the most appropriate way to represent the following set of data?

56	39	87	40	47	76	109	89	101	67	58	43
79	105	46	63	102	99	41	65	104	83	77	101
92	95	49	52	80	88	57	67	104	69	57	40
59	62	89	53	71	85	91	110	94	42	81	68
84	78	93.									

a) univariate scatterplot

b) frequency histogram

c) frequency polygon

d) grouped-frequency histogram

e) grouped-frequency polygon.

9. The following is a set of error scores on a 500 item multiple-choice test:

172	47	19	129	62	234	41	137	164	11	230
89	74	275	130	0	320	22	1	145	44	89
200	219	3	95	153	14	195	101	25	65	71
246	117	188	25	128	115	33	161	66	9	81
183	14	143	39	158	207	155	163	173	222.	

You are asked to summarize these scores by a frequency distribution table. Would you group the data, and if yes, what interval width would you use?

a) no grouping

b) grouping, with interval width 10

c) grouping, with interval width 20

d) grouping with interval width 50

e) grouping, with interval width 100.

10. The following univariate scatterplot represents the distribution of heights in a sample of 15 nine-year-old boys:

```
-|----X-XXX|X-XXX---X|XX-----X-|----X----|---X---X-|-

110        120        130        140        150        160
```

Height (cm)

What is the best description for this distribution?

a) symmetric

b) idealized

c) uniform

d) skewed-to-the-right

e) skewed-to-the-left.

Analysis

Correct Answer	Explanation of Wrong Answers
1b	a c see Study Guide Tip A: the smallest possible difference for scores recorded to one decimal place is 0.1, not 1.0 or 0.01.
	d e review the definition of real limits, Text Section 2.4.
2b	a you calculated the correct width; to find the midpoint, add half the width to the lower real, not apparent, limit of the interval.
	c d e see Study Guide Tip A: the width of an interval is the distance between its real limits.
3b	a you should subtract half of the width from the midpoint to obtain the lower real limit.
	c this is the lower apparent limit of the interval.
	d this is the upper real limit.
	e you should subtract half of the width from the midpoint.
4c	a b convert artists' factor 5 frequency into relative frequency or businessmen's factor 5 relative frequency into frequency, and compare the two groups' frequencies or relative frequencies.
	d essential information for conversion between frequency and relative frequency, i.e., sample size, is provided.

Correct Answer	Explanation of Wrong Answers
5b	a c convert relative frequencies of score 53 into frequencies for A and B, and compare.
	d essential information for conversion between frequency and relative frequency, i.e., sample size, is provided.
6e	a b the use of simple versus relative frequencies does not affect the shape of a distribution.
	c d the use of simple or relative frequencies is equivalent for comparison of data sets of equal sample size.
7a	b c d e reread Text Section 2.5 and Figure SG 2.1.
8d or e	a see Objective 11c.
	c b see Objective 9c.
9d	a see Objective 9c.
	b c e see Objective 9d.
10d	a b c e see Objectives 13 and 16.

CHAPTER 3. CENTRAL TENDENCY

OBJECTIVES

Chapter 2 introduced the frequency distribution as a way to represent and summarize a set of data. Chapter 3 discusses how a distribution can be further summarized by a measure of its center. You should be familiar with

I the three commonly used measures of central tendency: the mean, median and mode.

Since you start doing your own calculations in this chapter, you should also be familiar with

II the principles of reasonableness and usefulness, which you will have to apply to the results of any calculation.

More specifically, after reading the text you should be able to

I K (1) define the terms <u>mean</u>, <u>median</u>, <u>mode</u>, <u>modal interval</u> (Text 3.2, 3.3, 3.4,);

A (2) compute the mean of a set of raw scores or a frequency distribution (Text 3.1);

(3) rank a set of scores (Text 3.3);

(4) determine the median for data sets with odd or even sample sizes (Text 3.3);

(5) determine the median of a set of scores given by a frequency distribution (Text 3.3);

(6) determine the mode(s) or modal interval(s) of a distribution of scores (Text 3.4);

II K (7) define the term <u>deviation</u> (Text 3.2);

(8) pronounce and define the following symbols: $\underline{y}$, $\underline{n}$, $\underline{y}_n$, $\bar{\underline{y}}$, Σ (Text 3.2; Study Guide Tip B);

(9) translate a rule or formula from sentence form into symbols and vice versa (Text 3.2; Study Guide Tip B);

C (10) interpret the mean as the balance point of a distribution, that is, as the point from which the sum of the deviations of all scores is zero (scatterplot as seesaw) (Text 3.2, 3.5);

(11) understand the relationship between the shape of a distribution and the order of values of the three measures of central tendency (Text 3.5; Figure SG 3.1);

A (12) apply the principle of fairness of representation (Chapter 2) to decide on the most appropriate measure of central tendency to summarize and describe a given set of data (Text 3.2, 3.5);

(13) decide on how to check the reasonableness of a particular calculation, i.e., how to obtain a rough estimate of the expected result (Text 3.6);

Figure SG 3.1 Relationships among the Three Measures of Central Tendency, Depending on the Shape of the Distribution

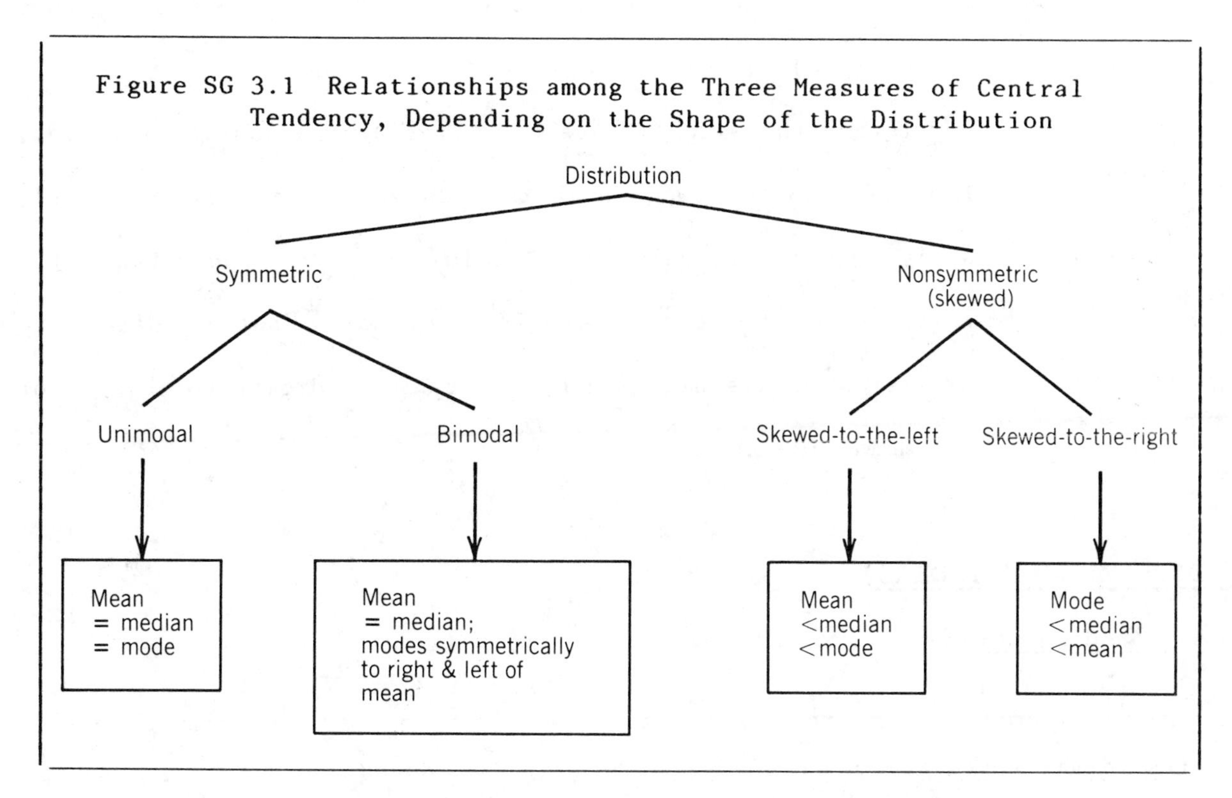

(14) make use of three properties of the mean when judging the reasonableness of a mean that you have calculated for a set of scores:

(a) the mean always has a value between the largest and smallest score values of the set (Text 3.6,);

(b) the value of the mean is usually similar to that of the median (Text 3.5, 3.6);

(c) depending on the skewness of the distribution, the mean is greater, smaller, or equal to the median (Text 3.5, 3.6);

(15) round a value to any desired number of digits (Text 3.6; Study Guide Tip A);

(16) understand the significance of 0 (zero) as a last digit, i.e., the difference between numbers such as 5 and 5.0 (Text 3.6);

(17) apply the principle of usefulness to understand the relationship between the sample size of a set of data, the precision of its mean, and the number of digits to retain for the mean (Text 3.6).

TIPS AND REMINDERS

A. Rounding Rules

First digit to be dropped	Round
0 - 4	down
6 - 9	up
5, followed by anything except 000 . . .	up
5, followed by no digit or 000 . . .	down *or* up, such that last retained digit becomes an even number (0, 2, 4, 6, 8)

If, by rounding, the last digit to be reported becomes 0 (zero), always report it (76.0 is not the same as 76).

Following these rules, round the following values to the indicated number of decimal places. The correct answers are listed at the end of the chapter.

	Number of Decimal Places	
16.53	1	
7.957	2	
0.55	1	
101.04	1	
0.7445	3	
43.97	1	
2.6351	2	
11.5	0	

B. Symbol Exercise

Match each symbol or equation in the left column with the expression that best describes it in the right column. The solutions are listed at the end of the chapter.

____ 1.	n	a.	y-bar
____ 2.	y_4	b.	distance of a score from the mean
____ 3.	Σ	c.	the mean is the sum of ys divided by n
____ 4.	$(y - \bar{y})$	d.	number of scores
____ 5.	$\bar{y}$	e.	the sum of the deviations of all scores from the mean is zero
____ 6.	$\Sigma(y - \bar{y}) = 0$	f.	y sub four
____ 7.	$\bar{y} = \frac{\Sigma y}{n}$	g.	sigma

PRACTICE QUIZ

Questions

1. A clinical psychologist measured the number of days during which 63 hospitalized psychiatric patients ingested a particular medication before they could be discharged into the community. He calculated the mean of these 63 integer scores to be precisely 18.9547. Which of the following values of the mean should he report in his publication of the results?

   a) 18
   b) 19
   c) 18.9
   d) 19.0
   e) 18.955.

2. The wage negotiations between a large company and its labor union have come to an impasse. An arbitrator has been appointed to resolve the dispute. The great majority of union members are low or medium wage earners; only a few specialists receive top wages. Which of the following measures of central tendency would you expect the union to submit to the arbitrator as a summary of its membership's current wages in order to influence him in the union's favor?

   a) mean b) median c) mode.

3. At a busy city intersection traffic jams occur every day. The local police department is understaffed so that it can provide a policeman to regulate the traffic at this intersection only for one hour per day. A survey of the number of cars that cross the intersection in an hour is taken for the different hours of the day. Which of the following measures of central tendency of this distribution of the number of cars per hour depending on the hour of the day is most relevant to the decision of when to station the policeman at the intersection?

   a) mean b) median c) mode.

4. Which score is most likely to be the median of a distribution skewed to the left with a mean of 57 and a mode of 65?

   a) 50 d) 65

   b) 57 e) 73.

   c) 60

5. As the difference between the mean and the median approaches zero, the shape of the distribution

   a) becomes more symmetric.

   b) becomes more skewed-to-the-left.

   c) becomes more skewed-to-the-right.

   d) becomes more uniform.

   e) is not affected.

6. Construct a grouped-frequency distribution for the following set of data, using 40-49 as the first interval.

50	88	66	91	43	101	96	63	78	81	70	41
84	73	93	55	74	61	76	64	85	58	107	74
73	89	115	97	79	84	49	83	69	54	113	67
94	103	68	75	86	103.						

What is the modal interval of this distribution?

   a) 8

   b) 9

   c) 60-69

   d) 70-79

   e) 80-89.

7. What is the median (to the nearest integer) of the scores summarized by the following frequency distribution?

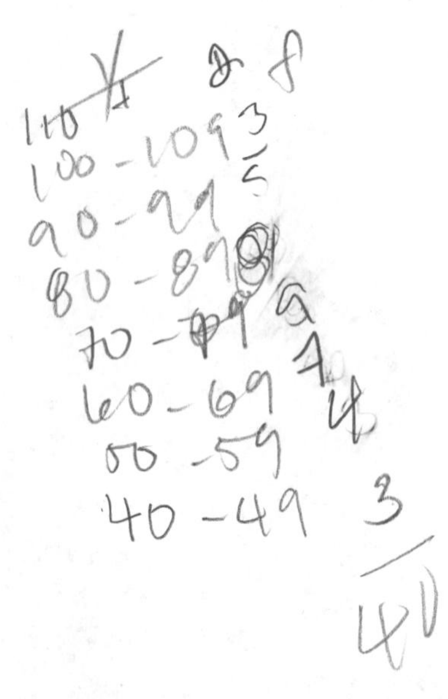

Score	f
11	7
12	12
13	11
14	0
15	8
16	12
17	6
18	2
19	1
20	0
21	1
	60

a) 13
b) 14
c) 15
d) 16
e) 30.

8. The following measurements represent the grade-point averages of 25 college students:

7.1 6.9 8.0 7.5 7.7 8.3 8.4 7.4 8.1 7.7 7.5
7.3 6.4 7.8 7.7 7.8 7.2 8.0 6.9 7.3 8.4 7.9
7.5 7.9 7.5.

What is the median grade-point average?

a) 7.7
b) 7.6
c) 7.5
d) 7.4.

9. What is the mean of the following frequency distribution?

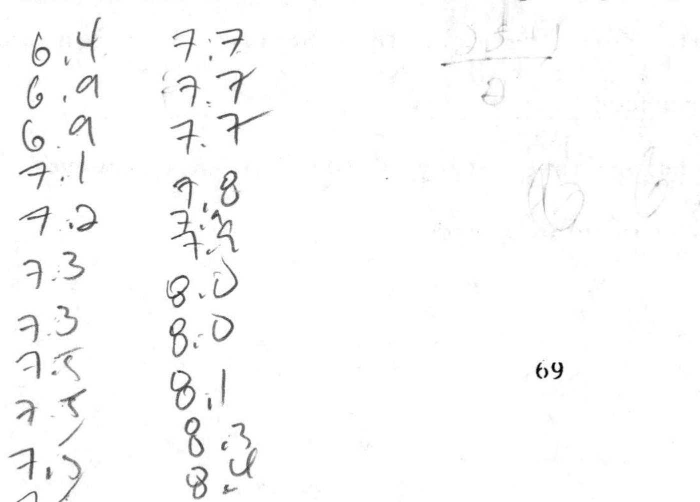

Score	Frequency
7	2
8	0
9	6
10	9
11	9
12	5
13	3

a) 4.9

b) 10.0

c) 10.2

d) 10.5

e) 10.7.

10. Thirty students write the first take of the statistics midterm exam. Their mean exam score is 70. The remaining 20 students write the second take of the exam and obtain a mean score of 60. What is the mean exam score for all 50 students?

a) 60

b) 65

c) 66

d) 70

e) 110.

Analysis

Correct Answer	Explanation of Wrong Answers
1d	a b you should report the mean to one more decimal place than the original scores (reread Text Section 3.6); in answer a you also rounded incorrectly.
	c incorrect rounding (see Study Guide Tip A); however, the precision of the mean is correct.

Chapter 3 CENTRAL TENDENCY

Correct Answer	Explanation of Wrong Answers
	e this value of the mean is *not useful*; the last two digits are "noise"; reread Text Section 3.6.
2b or c	a see Objectives 11 and 12.
3c	a b see Objectives 11 and 12.
4c	a b d e reread Text Section 3.5 and Study Guide Figure 3.1.
5a	b c d e reread Text Section 3.5 and Study Guide Figure 3.1.
6d	a b are *frequencies* rather than intervals; answer b is the *frequency* of the *modal interval*, answer a is the *frequency* of the second most frequent interval.
	c e check your grouped-frequency distribution: there probably is a mistake in your frequency counts.
7b	a c for *even* $\underline{n}$, the median is *not* the score with rank $\underline{n}/2$ (answer a) nor with rank ($\underline{n}/2 + 1$) (answer c), *but* the *mean of those two scores*.
	d you calculated the median of the scores listed in the frequency distribution *as if* each of them occurred only *once*; reread Text Section 3.3.
	e this is the *rank* of the median (rounded to the nearest integer); you still have to find the *score* which corresponds to the rank of 30.5.

Correct Answer	Explanation of Wrong Answers
8a	b c for *odd* n, the median is *not* the score with rank $n/2$ (answer b) nor with rank ($n/2$ - 1/2) (answer c), *but* the score with rank ($n/2$ + 1/2).
	d the median is *not* the average of the highest and the lowest score of the data set.
9d	a this is the *mean frequency*, *not* the mean of the distribution.
	b you calculated the mean of the scores listed in the frequency distribution *as if* each of them occurred only *once*; reread Text Section 3.1.
	c e the product (8 × 0) is zero; thus, a score with frequency 0 is neither added to (answer e) nor subtracted from Σy (answer c).
10c	a b d e the *sum* of the scores of the 30 students must be 30 × 70 = 2100; the *sum* of the scores of the 20 students must be 20 × 60 = 1200; hence the sum of all 50 scores is 2100 + 1200 = 3300 and the mean is 3300/50 = 66.

Rounding Solutions: 16.5, 7.96, 0.6, 101.0, 0.744, 44.0, 2.64, 12.

Symbol Exercise Solutions: 1d, 2f, 3g, 4b, 5a, 6e, 7c.

CHAPTER 4. VARIABILITY

OBJECTIVES

Chapter 4 concludes the discussion which was started in Chapters 2 and 3 of how to summarize and describe a set of data. A measure of the variability of a distribution of scores, which is introduced in Chapter 4, together with a measure of its central tendency, can describe any distribution fairly accurately (general objective II). You should be familiar with

II the most commonly used measures of variability: standard deviation, variance, interquartile range, and 68% range.

The last two measures of variability are defined in terms of percentiles, which should be studied first (general objective I). You should be familiar with

I the concept and the calculation of percentiles.

More specifically, after reading the text you should be able to

I K (1) define the terms <u>percentile</u>, <u>cumulative frequency</u>, <u>forward problem</u>, <u>backward problem</u>, and <u>quartile</u> (Text 4.1);

(2) express the relationship between cumulative frequency below midpoint, sample size, and percentile by a formula (Text 4.1);

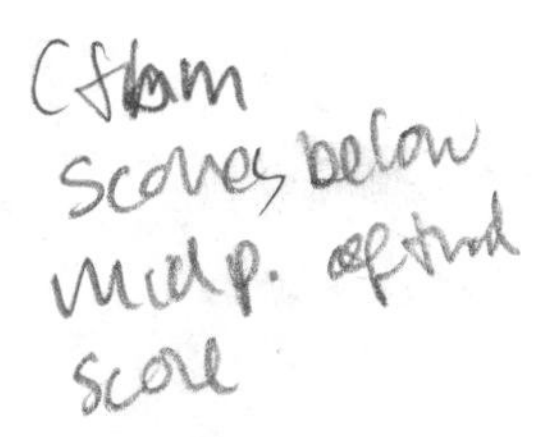

C (3) identify and distinguish between a forward and a backward problem (Text 4.1);

(4) interpret a percentile graphically as the percentage of the area of a frequency distribution below a given score (Text 4.1; Study Guide Tip A);

(5) estimate the reasonableness of a calculated score or percentile from its graphic representation (Text 4.1);

(6) apply the principle of usefulness when calculating percentiles or corresponding scores (Text 4.1);

A (7) compute the percentile of a given score from a set of scores or from a frequency distribution (Text 4.1);

(8) compute the score that is at a given percentile in a distribution (Text 4.1);

II K (9) define the terms variability, range, interquartile range, 68% range, standard deviation, variance, and variability principle (Text 4.2, 4.3, 4.4)

C (10) evaluate the advantages and disadvantages of the various measures of variability (Text 4.2);

(11) distinguish between the variability and the central tendency of a distribution (Text 4.2);

(12) give a rough estimate of the 68% range of a distribution (from its graph) (Text 4.2);

(13) identify the conditions under which the variability principle applies (Text 4.3);

(14) estimate the standard deviation of a distribution by using the variability principle (Text 4.3);

(15) evaluate the advantages and disadvantages of the definitional and the computational formulas for the standard deviation and variance (Text 4.4);

(16) interpret the information provided by a standard deviation or variance (Text 4.4);

A (17) use either the definitional or computational formula to calculate the variance or standard deviation of a set of data (Text 4.4; Study Guide Tip B);

(18) check any calculated standard deviation or variance for its reasonableness by comparing it to an estimated value (see Objectives 12 and 14) (Text 4.2, 4.3, 4.4; Study Guide Tip B).

TIPS AND REMINDERS

A. Graphic Representation of Percentiles

Graphically, a percentile translates into the percentage of the area of a distribution which lies below a particular score. The median as the 50th percentile is the score which divides the area of the distribution in half. Quartiles are the scores which divide a distribution of scores into quarters.

Def $\sqrt{\frac{\Sigma(y-\bar{y})^2}{n}}$ Comp $\sqrt{\frac{\Sigma y^2-(\Sigma y)^2/n}{n}}$

Chapter 4 VARIABILITY

In a symmetric distribution the mean and the median coincide, thus putting the mean at the 50th percentile. Due to the symmetry of the distribution, the distance from the median (= second quartile) to the first quartile and to the third quartile is the same (the actual distance will depend on the variability of the scores). In a skewed distribution however, the bulk of the distribution (i.e. area) lies in one of the tails. To counteract this "weight", the mean as the balance point moves away from the median into the other tail of the distribution (Chapter 3).

Figure SG 4.1 Quartiles in a Skewed Distribution

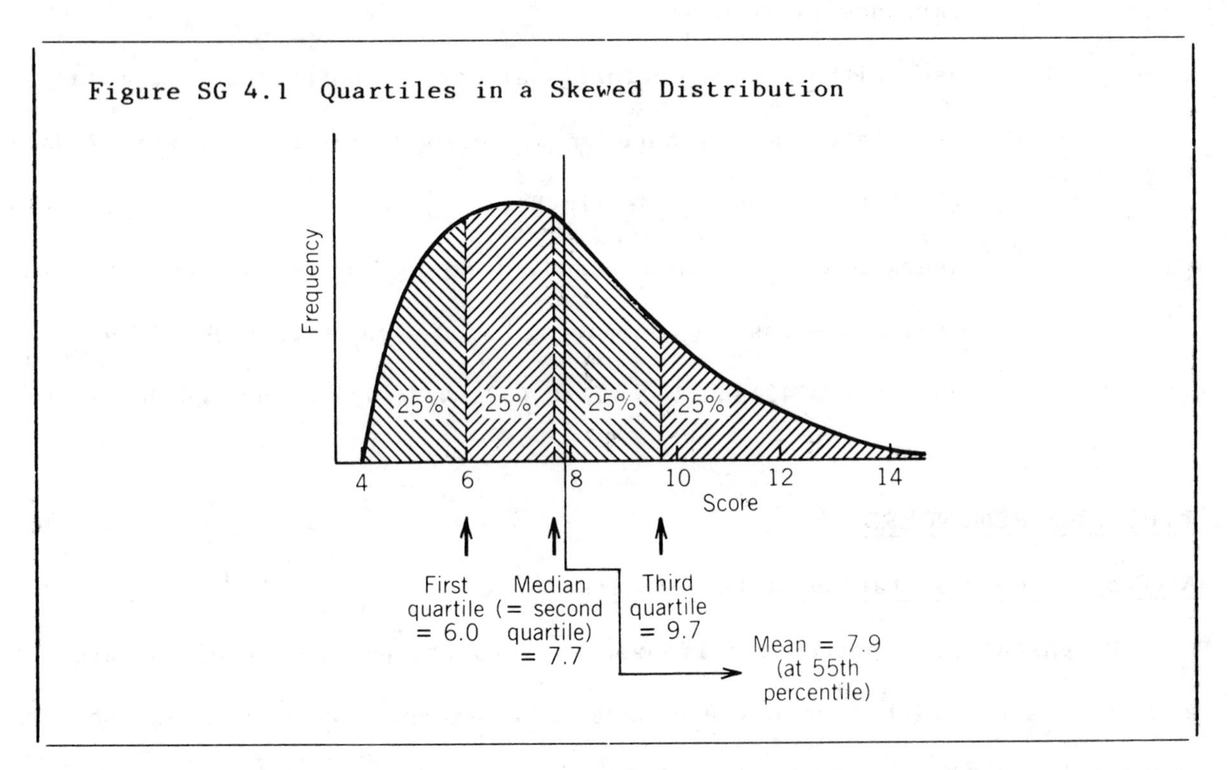

Figure SG 4.1 shows a positively-skewed distribution of a continuous variable. Note that the mean is approximately at the 55th percentile because of the reason given at the end of the last paragraph. Also, due to the skewness, the distance between the median and the first quartile is smaller than the distance between the median and the third quartile, though these two intervals, by definition, contain equal numbers of scores. By just comparing these two interquartile distances, some general predictions about the shape of the distribution can be made.

B. Calculating a Standard Deviation

There are four steps involved in calculating a standard deviation:

- compute the standard deviation using the appropriate formula;
- round the computed value to a useful number of digits;
- estimate the standard deviation by the variability principle;
- compare the computed and estimated values; in case of a discrepancy, repeat the computation.

Choosing the appropriate formula. You have to decide between the definitional formula and the computational formula. Remember that the formulas are equivalent and will lead to the same result. Their difference lies in the ease of computation. Unless a distribution consists of only a few scores which have a "nice" mean, the computational formula for s' is commonly used. Computing and squaring the deviation of each score from the mean as required by the definitional formula is a ready source of error for most sets of data.

Mechanics of Calculation. To calculate a standard deviation using the computational formula, you need to know the values of the sum of all scores

(Σy) and the sum of the squares of all scores (Σy^2). To obtain Σy you simply add up all the scores; to obtain Σy^2 you have to square each score before you add the squares. To avoid computational errors you should not copy these squared scores down. Your calculator probably has a "memory". As you compute each square you can add the square to the memory. When you have finished computing all the squares, the memory will contain the required sum.

C. Symbol Exercise

Match the symbol or equation in the left column with the expression in the right column which best describes it. The solutions are at the end of the chapter. Note: CFBM means cumulative frequency below midpoint.

c 1.	s'	a. s'^2
a 2.	$\dfrac{\Sigma(y - \bar{y})^2}{n}$	b. CFBM
e 3.	square root of the sum of squared deviations divided by n	c. $\sqrt{\dfrac{\Sigma(y - \bar{y})^2}{n}}$
d 4.	$\dfrac{CFBM}{n} \times 100$	d. percentile
b 5.	$n \times \dfrac{percentile}{100}$	e. s'

PRACTICE QUIZ

Questions

1. An applicant's score is at the 90th percentile on an aptitude test. This means that she

   a) placed ninetieth in the group taking the test.

   b) had twice as many correct items as someone at the 45th percentile.

   c) performed as well as or better than 90% of the group taking the test.

   d) answered 90% of the total number of items correctly.

   e) none of the above.

2. In what type of distribution might the mean be at the 60th percentile?

   a) skewed-to-the-right

   b) skewed-to-the-left

   c) symmetric

   d) bimodal

   e) uniform.

3. Which of the following is the percentile (rounded to the nearest integer) of a score of 40 as computed from this frequency distribution?

Score	f
37	5
38	13
39	29
40	34
41	31
42	18
43	7
44	4
	141

a) 67

b) 64

c) 57

d) 55

e) 45.

4. What score (rounded to one decimal place) is at the 39th percentile of the following data set ($\underline{n}$ = 43)?

3.3	3.4	3.6	3.7	3.7	3.7	3.8	4.0	4.3
4.3	4.4	4.5	4.5	4.6	4.7	4.7	4.8	4.9
5.1	5.1	5.4	5.5	5.6	5.9	6.0	6.0	6.1
6.5	6.7	7.0	7.2	7.2	7.3	7.4	7.5	7.6
7.6	7.6	7.7	7.8	8.1	8.3	8.3		

a) 3.7

b) 4.8

c) 6.1

d) 7.8

e) 16.8.

5. Choose the best estimate for the 68% range of the following data set:

23	25	28	30	31	33	33	34	35
35	35	36	36	36	36	37	37	38
38	38	39	40	42	45			

a) 22

b) 16

c) 14

d) 8

e) 4.

6. Choose the best <u>estimate</u> for the standard deviation of the following frequency distribution (do <u>not</u> <u>calculate</u> the standard deviation):

Score	f	Score	f
8.6	1	9.5	3
8.7	2	9.6	3
8.8	0	9.7	0
8.9	4	9.8	2
9.0	7	9.9	1
9.1	9	10.0	3
9.2	13	10.1	1
9.3	8		
9.4	5		62

a) 0.1
b) 0.3
c) 1.0
d) 3
e) 10.

7. An experimenter tabulates the number of errors made on a manual dexterity test by 50 subjects. The best subject made only 7 errors, the poorest 28 errors. Some percentiles of the distribution of number of errors were:

Percentile	16th	32nd	68th	84th
Number of errors	13	14	19	21

Which of the following is the best estimate of the standard deviation for this distribution?

a) 4
b) 6
c) 8
d) 10.5
e) 12.

8. A group of eleven students took a spelling test; the distribution of their marks turned out to have a mean of 8 and a standard deviation of 1. What were the eleven spelling marks obtained?

a) all marks are 7

b) all marks are 8

c) all marks are 9

d) impossible to say without further information.

9. Compute $\underline{s}'$ for the following set of scores:

10 9 8 6 3 2 1

a) 3.3

b) 3.6

c) 8.8

d) 11.1

e) 77.7.

10. Compute $\underline{s}'^2$ for the following set of scores:

.16	.12	.10	.14
.13	.15	.18	.13

a) .0005

b) .0006

c) .0043

d) .0231

e) .0655.

Analysis

Correct Answer	Explanation of Wrong Answers
1c	a b d e review the definition of percentile.
2a	b c d e see Text 4.1 and Study Guide Tip A.

Correct Answer	Explanation of Wrong Answers
3e	a c you should use the cumulative frequency below the midpoint of the score interval 40, not the cumulative frequency which includes the whole interval; for answer c, in addition, the frequencies should be cumulated starting with the lowest score regardless of whether the scores are listed in increasing or decreasing order.
	b is an intermediate result, i.e. the cumulative frequency below the midpoint of the score interval 40, not the percentile.
	d you should always cumulate the frequencies starting with the lowest score regardless of whether the scores are listed in increasing or decreasing order.
4b	a d you omitted the step of computing the rank (cumulative frequency below midpoint) which corresponds to the 39th percentile; instead, you used the percentile as if it were the rank; also, for answer a, always rank a set of scores in increasing order.
	c always rank a set of scores in increasing, not decreasing, order.
	e is an intermediate result, i.e. the rank (cumulative frequency below midpoint) of the score which is at the 39th percentile, not the score itself.

Chapter 4 VARIABILITY

Correct Answer	Explanation of Wrong Answers
5d	a the full range is too large an estimate for the 68% range.
	b is an intermediate result, i.e., the number of scores included in the 68% range, not the 68% range itself.
	c is the range of the lower 68% of the scores, not the 68% range which is defined as the interval including the middle 68% of the scores.
	e is too small an estimate for the 68% range.
6b	a c d e according to the variability principle, the standard deviation is approximately one-half the 68% range; the 68% range extends from about 9.0 to about 9.6, a distance of about 0.6. Half of this is 0.3.
7a	b is one-half of the range of the lower 68% of the scores, which is not the 68% range (see answer 5c).
	c for an estimate of the standard deviation, you should divide the 68% range in half.
	d is one-half of the full range, not of the 68% range.
	e is the range of the lower 68% of the scores, not the 68% range (see answer 5c). You also did not divide the range in half.

Correct Answer	Explanation of Wrong Answers
8d	a b c in these cases the standard deviation would be 0; see the definitional formula in Text 4.4.
9a	b is the adjusted standard deviation, s (Chapter 11).
	c is the squareroot of the sum of the squared deviations, i.e. the numerator of the standard deviation formula.
	d is the unadjusted variance.
	e is the sum of the squared deviations.
10a	b is the adjusted variance, s^2 (Chapter 11).
	c is the sum of the squared deviations, i.e. the numerator of the variance formula.
	d is the standard deviation.
	e is the squareroot of the sum of square deviations.

Symbol Exercise Solutions

1c, 2a, 3e, 4d, 5b.

REVIEW CHAPTER A

OBJECTIVES

Chapter A is the first review chapter in the text. One way to review Chapters 2, 3, and 4 is to summarize them in tree diagrams. You should be familiar with

I tree diagrams.

You should review

II Chapters 2, 3, and 4.

More specifically, after reading the text you should be able to

I K (1) define the parts of a tree diagram: stage, branch, and path (Text A.3);

C (2) distinguish between decision trees and summary trees (Text A.1, A.2);

(3) distinguish between tree diagrams which have the same set of branches for each branch of the preceding stage and tree diagrams which have different sets of branches for each branch of the preceding stage (Text A.3);

Chapter A REVIEW

A	(4)	construct the most useful tree diagram to represent a given situation (Text A.3);
II C	(5)	understand the relationship between topics in Chapters 2, 3, and 4 (Text A.2).

Note: There are no "tips and reminders" or "practice quiz" for this chapter. You should review Chapters 2, 3, and 4 of the text and study guide.

CHAPTER 5. RANDOM SAMPLING AND PROBABILITY

OBJECTIVES

Chapter 5 introduces the theory of probability. You should be familiar with

I the concept of random sampling;

II the theoretical and empirical definition of probability;

III the binomial distribution.

You will find the tree diagram techniques introduced in Chapter A useful in solving many probability problems.

More specifically, after reading the text you should be able to

I K (1) define the terms population, sample, sample space, full sample space, reduced sample space, probability (theoretical definition), random sample (Text 5.1, 5.2, 5.3);

C (2) distinguish between the tree diagrams if the variable is discrete vs. continuous (Text 5.2);

A (3) represent the sample space of samples of size n from a given population by a tree diagram (Text 5.2);

(4) determine the number of samples in a sample space arithmetically or with help of a tree diagram (Text 5.2);

(5) compute the probability of a given sample from the branch probabilities of its path through the sample space tree (Text 5.3);

(6) reduce the size of a tree diagram by assigning one branch to each different value of the variable under study (Text 5.2,);

II K (7) define the terms *probability* (*empirical definition*), *law of large numbers*, *random digit*, *fair coin*, *biased coin*, *sampling with replacement*, *sampling without replacement* (Text 5.4, 5.5);

(8) use the random number table (Text 5.4);

C (9) understand the relationship between the theoretical and empirical definition of probability via the law of large numbers (Text 5.4);

(10) distinguish between a fair and biased coin or die (Text 5.5);

(11) distinguish between sampling with replacement and sampling without replacement (Text 5.5);

A (12) estimate the probability of a result from an observed relative frequency and determine the expected frequency in a sample of size *n* of a result which has a given probability (Text 5.4; Study Guide Tip A);

III K (13) define the terms probability distribution, binomial distribution, binomial tree, binomial probability, success, failure (Text 5.6);

A (14) represent the sample space for a given study (with appropriate characteristics) by a binomial tree (Text 5.6);

(15) determine the values of n, p, q for a study which can be represented by a binomial tree (Text 5.6);

(16) compute binomial probabilities from a) first principles (Text 5.6) or b) the binomial formula (Text 5.6; Study Guide Tip B).

TIPS AND REMINDERS

A. Probability and Expected Frequency

The expected relative frequency of a result, in a sample, can be predicted if its probability, p, is known:

expected relative frequency = p.

For a sample of size n, frequency and relative frequency are related:

frequency = n × relative frequency.

Hence, the expected frequency of a result in a sample of size n can be predicted from the probability of the result:

expected freqency = n × expected relative frequency

= n × p.

B. The Binomial Formula

As described in Text Section 5.6, the formula for computing a binomial probability is the product of two terms: (a) the number of paths in a binomial tree with the desired number of successes, y, and (b) the probability of one such path with y successes:

$$P(y) = \frac{n!}{y!\,(n-y)!} \times p^y q^{n-y}.$$

Manipulating Factorials. A common mistake in part (a) of the formula (quotient of three factorial expressions) is incorrect cancellation of the factorials. If $n = 6$ and $y = 4$,

$$\frac{6!}{4!\,(6-4)!} = \frac{6!}{4!\,2!} = \frac{6\times5\times4\times3\times2\times1}{(4\times3\times2\times1)(2\times1)} = \frac{6\times5}{2} = 3\times5 = 15.$$

It is incorrect to cancel $6!/(4!\ 2!)$ to $3/4$, thus ignoring the factorial symbol.

Manipulating Powers. A common mistake in part (b) of the formula (the product of two powers) is to interchange the powers y and $(n - y)$ to which p and q are raised, respectively. For $p = .3$, $q = .7$, $n = 3$, and $y = 2$, the expression

$$p^y q^{n-y} = (.3)^2(.7)^1 = (.3)(.3)(.7)$$

is not equivalent to

$$p^{n-y} q^y = (.3)^1(.7)^2 = (.3)(.7)(.7).$$

Unless $p = q = .5$ where the two expressions both simplify to $p^y q^{n-y} = p^{(y+n-y)} = p^n$, interchanging the powers of p and q in the binomial formula will lead to an incorrect answer.

PRACTICE QUIZ

Questions

1. Refer to the tree diagram in Figure SG 5.1. Which of the following experiments is illustrated by this tree diagram?

Figure SG 5.1 Practice Quiz, Question 1

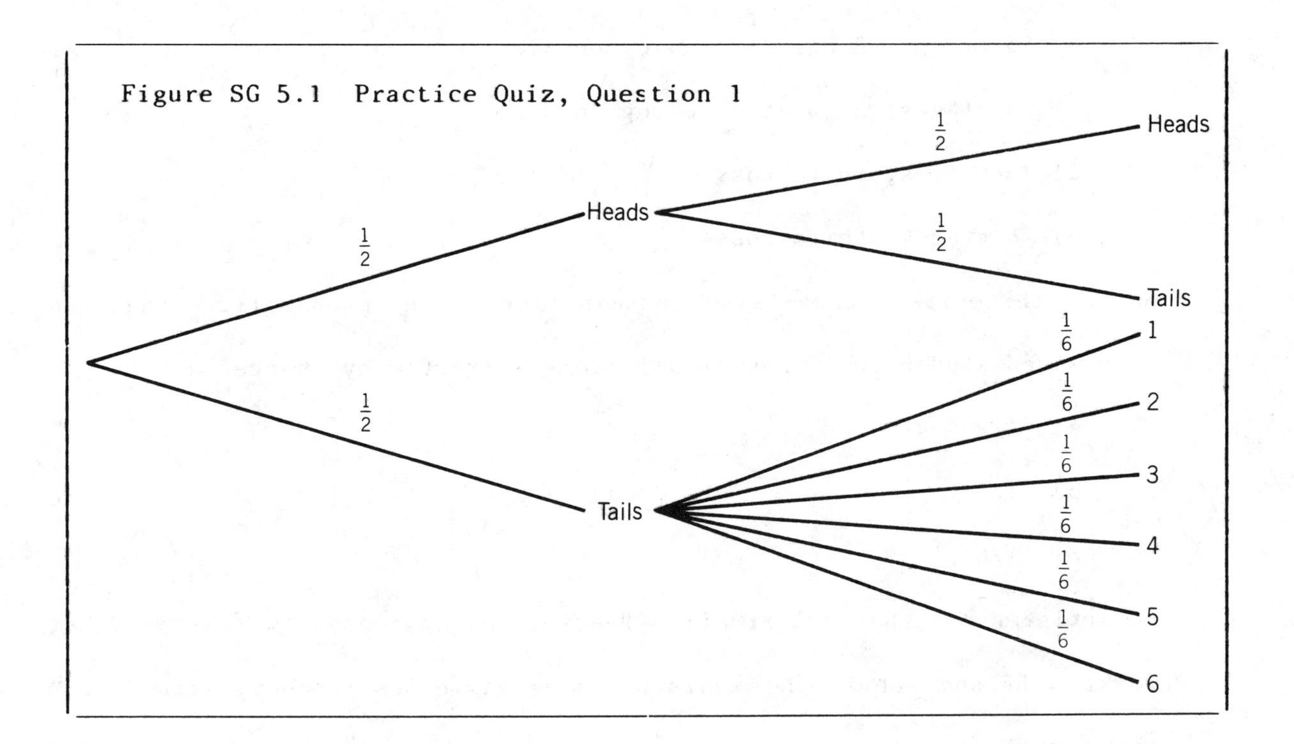

a) Two coins are tossed and one die is rolled and the results are recorded.

b) A coin is tossed. If heads results, the coin is tossed again; if tails results on the first toss, a die is rolled.

c) A coin is tossed twice. If tails results, a die is rolled, and the results are recorded.

d) A coin is tossed and a die is rolled and the results are recorded.

e) A coin is tossed. If heads results, the coin is tossed again; if tails results on the first toss, six dice are rolled.

2. A die is tossed three times. The possible results are represented in a tree diagram. How many stages does this diagram have, and what is represented by the 3rd stage?

a) 6 stages; 3 points showing on third toss

b) 6 stages; 3 points showing on die

c) 3 stages; 3 points showing on die

d) 3 stages; third toss

e) 6 stages; third toss.

3. In a three-item true-false examination, the probability that an unprepared student will answer all items correctly by chance is

a) 1/2

b) 1/3

c) 1/6

d) 1/8

e) 1/27.

4. An artist knows that 70% of all galleries she approaches will reject her works. If she submits her work to three galleries randomly picked from the phonebook, what is the probability that all three will reject it?

a) .900

b) .343

c) .233

d) .027

e) .000.

5. Consider an experiment consisting of drawing three marbles from a bag containing a very large number of green and orange marbles (equal numbers of each). What is the number of different samples for this experiment?

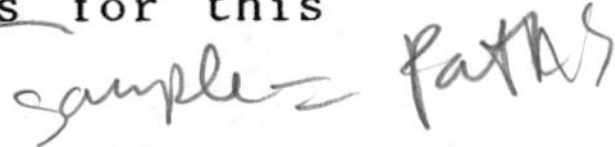

a) 27

b) 8

c) 2

d) 3

e) impossible to say, because the number of marbles in the bag is not known.

6. In a large department store the chance of a particular type of article being stolen is 1 in 100. What is the probability that in two consecutive cases of shoplifting it is this type of article that was stolen?

a) .0001

b) .0050

c) .0200

d) .4950

e) .9801.

7. A U.N. committee consists of seven members from three countries: two Nigerian, three Chinese, and two French. A major and a minor delegate are to be chosen to represent the committee in a larger assembly.

Draw tree diagrams to represent the two situations that i) each of the members is eligible to become either the major or minor delegate (full sample space); ii) only the nationality of the members is of importance, that is, the major and minor delegates must be from different countries and it is immaterial which of the particular members is representing his or her country (reduced sample space). How many samples are in each of the two sample spaces?

a) full: 49; reduced: 9

b) full: 49; reduced: 6

c) full: 42; reduced: 6

d) full: 42; reduced: 42

e) full: 14; reduced: 6.

8. An experiment consists of 1000 trials, each of which has 16 equally probable outcomes. At the end of the experiment, what would be the approximate frequency of each outcome?

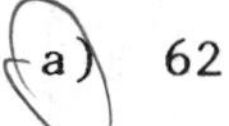

a) 62

b) 1

c) 1/16

d) 1/1000

e) impossible to say without further information.

9. In a particular city, 1/3 of the population has bloodtype B. What is the probability that exactly one of four randomly selected individuals will have bloodtype B?

a) .083

b) .099

c) .132

d) .333

e) .395.

10. Two dice are rolled. Construct the distribution of the statistic y = total number of points showing on both dice. (Tip: Work from first principles!) What is the probability of $y = 8$?

a) 4 d) 1/36

b) 5 e) 4/30

c) 1/30 f) 5/36.

Analysis

Correct Answer	Explanation of Wrong Answers
1b	a c d e you should review the significance of the stages and branches of a tree diagram in representing an experiment (Text Chapter A; Text 5.2).
2d	a b c e see Answer 1 a c d e
3d	a this is the branch probability of obtaining a correct answer by chance, not the probability of the result of 3 correct answers in a sample of size 3;
	c e you probably constructed an incorrect tree diagram with branch probabilities of 1/3;
	b this probably is the branch probability of the incorrect tree diagram of answers b or e.
4b	a e you probably failed to draw a tree diagram to represent the situation.
	c you probably constructed an incorrect tree diagram;

Correct Answer	Explanation of Wrong Answers
	d this is the wrong path through the tree, i.e., the probability of three acceptances.
5b	a you probably constructed an incorrect tree diagram with 3 branches at each stage;
	c this is the number of branches at each stage, not the number of different samples;
	d this is the number of branches at each stage for the incorrect tree diagram of answer a;
	e the only relevant information for determining the number of different samples of an experiment is the number of different choices (i.e., branches) at each stage.
6a	b c d you probably failed to draw a tree diagram to represent the situation;
	e is the probability that the article is not stolen in two consecutive thefts.
7c	a is the number of samples for the full and reduced sample space if sampling was with replacement;
	b is the number of samples for the full sample space if sampling was with replacement; the number of samples for the reduced sample space is correct;
	d e see answer 6 b.

Correct Answer		Explanation of Wrong Answers
8a 	b	is the frequency of each trial;
	c	is the relative frequency of each outcome (see Objective 12);
	d	is the relative frequency of each trial;
	e	see Objective 12; Study Guide Tip A.
9e	a	you have to use the binomial formula;
	b	is the probability of one path with $y = 1$, not all paths;
	c	see Study Guide Tip B, Manipulating Factorials;
	d	is p, the probability of bloodtype B, not the probability of $y = 1$.
10f	a	is the number of samples for which $y = 8$, for a tree diagram which incorrectly assumes "sampling without replacement";
	b	is the number of samples for which $y = 8$;
	c	is the probability of any one sample for the incorrect tree diagram of answer a;
	d	is the probability of any one sample;
	e	is the probability that $y = 8$ for the incorrect tree diagram of answer a.

CHAPTER 6

STATISTICAL INFERENCE AND THE DISTRIBUTION OF THE STATISTIC

OBJECTIVES

Chapter 6 continues the discussion of statistical inference started in Chapter 5. You should be familiar with

I the structure and purpose of statistical inference;

II the important concept of the distribution of a statistic--a probability distribution which relates a population parameter and a sample statistic;

III the binomial distribution as a particular type of a distribution of a statistic;

IV the two techniques of statistical inference: hypothesis testing and estimation.

More specifically, after reading the text you should be able to

I K (1) define the terms descriptive statistics, inferential statistics, population, sample, survey, experiment, sample space, random sample, representative sample, sample distribution, probability, population distribution, probability distribution (Text 6.1);

C (2) distinguish between a survey and and experiment (Text 6.1);

(3) state the purpose of statistical inference (Text 6.1);

(4) list some criteria which influence the accuracy of an inference from a sample to a population (Text 6.1);

(5) describe the role of the sample space in statistical inference making (Text 6.1);

(6) describe the importance of random sampling for statistical inference (Text 6.1);

(7) distinguish between sample distributions (i.e. relative frequency distributions) and population distributions (i.e. probability distributions) (Text 6.1).

II K (8) define the terms population parameter, sample statistic, distribution of a statistic (Text 6.2, 6.3);

(9) list some possible population parameters and the corresponding sample statistics (Text 6.2);

III K (10) define the terms discrete random variable, continuous random variable, discrete probability distribution, continuous probability distribution, mean of a random variable, standard deviation of a random variable, population mean, population standard deviation, Bernoulli distribution, binomial distribution, success, failure (Text 6.3, 6.5, 6.6);

(11) pronounce and identify the following symbols: p, q, μ, σ, σ^2, y, x (Text 6.3, 6.5, 6.6; Study Guide Tip C);

C (12) distinguish the two types of probability distributions: population distributions and distributions of a statistic (Text 6.5, Figure SG 6.1);

(13) interpret the binomial distribution as a particular case of a distribution of a statistic, and thus as a particular case of a (discrete) probability distribution (Text 6.3, 6.5; Figure SG 6.1);

Figure SG 6.1 Relationships among Various Probability Distributions

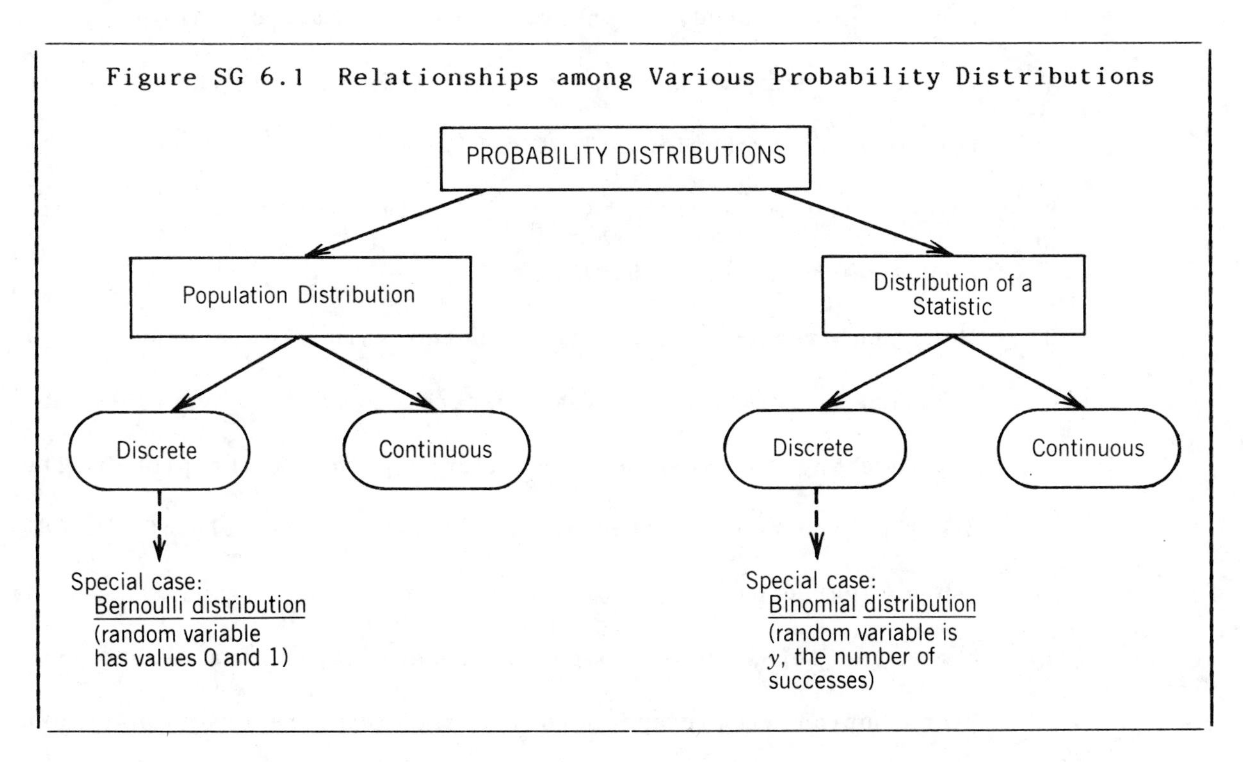

(14) decide whether the statistic in a given experiment or survey has a binomial distribution (Text 6.3);

(15) distinguish between the two types of probability distributions (random variables): the Bernoulli distribution (population distribution) and the binomial distribution (distribution of the statistic) (Text 6.6; Figure SG 6.1);

A (16) determine the values of $\underline{n}$, $\underline{p}$, $\underline{q}$ for a given study (Text 6.3);

(17) compute binomial probabilities from the binomial table (Text 6.3; Study Guide Tip A);

(18) for a given study, construct the population distribution (Bernoulli distribution) and the distribution of the statistic (binomial distribution) (Text 6.6);

(19) compute the mean and standard deviation of a given discrete random variable (Text 6.6);

(20) check the reasonableness of computed values of a mean and/or standard deviation of a discrete random variable by interpreting the mean as the balance point of the probability distribution and the standard deviation as one-half of the 68% range (Text 6.6);

(21) check the legality of computed probabilities of a probability distribution with respect to the following restrictions (Text 6.5):

a) all probabilities must be non-negative;

b) no single probability may exceed 1.0;

c) the sum of all probabilities must equal 1.0;

IV K (22) define the terms hypothesis testing, estimation (Text 6.4);

(23) state the basic principle of statistical inference (Text 6.4);

C (24) understand the role of the distribution of the statistic in statistical inference (Text 6.3, 6.4);

(25) distinguish between hypothesis testing and estimation (Text 6.4);

(26) describe how the basic principle of statistical inference applies to hypothesis testing and estimation (Text 6.4; Study Guide Tip B).

TIPS AND REMINDERS

A. Notation for Cumulative Probabilities

The probability of 8 heads in 10 tosses of a coin is represented by the symbol P(8). More accurately, P(8) stands for the probability of exactly 8 head in 10 tosses. In contrast, the probability of at least 8 heads in 10 tosses is the cumulative probability of obtaining 8 or more heads, i.e. P(8) + P(9) + P(10). Similarly, the probability of at most 3 heads in 10 tosses is the cumulative probability of obtaining 3 or less heads, i.e., P(0) + P(1) + P(2) + P(3).

B. Hypothesis Testing

The decision whether to accept or reject a hypothesis about the value of a population parameter is based on the probability of the observed value of the statistic in a random sample drawn from the population (Text Section

6.4). The relevant probability for deciding whether the observed statistic is likely or unlikely under the hypothesis is a cumulative probability, i.e., the probability of the observed value or a more extreme value (see Study Guide Tip A).

In order to decide which cumulative probability to compute, you should first compute the mean of the binomial distribution, $\mu = np$ (Text Section 6.6). This mean is the expected number of successes, y, if the hypothesis is true.

If the observed value is greater than the expected value of y, the relevant probability is the probability of at least the observed value. For example, if $p = 0.5$ the expected number of successes in 10 trials is $10 \times 0.5 = 5$; if the observed number of successes in a random sample is 8, the relevant probability is P(8) + P(9) + P(10).

If the observed value is less than the expected value, the relevant probability is the probability of at most the observed result. For example, in 10 trials the expected number of successes is $10 \times 0.5 = 5$; if the observed number of successes in a random sample is 3, the relevant probability is P(3) + P(2) + P(1) + P(0).

C. Symbol Exercise

Match each symbol or equation in the left column with the expression that best describes it in the right column. The solutions are listed at the end of the chapter.

_____ 1.	p	a.	the number of successes in a binomial study.
_____ 2.	q	b.	σ^2
_____ 3.	y	c.	the standard deviation of a random variable.
_____ 4.	$\bar{y} = \frac{\Sigma y}{n}$	d.	$\Sigma y P(y)$
_____ 5.	σ	e.	the probability of success.
_____ 6.	$\Sigma(y - \mu)^2 P(y)$	f.	The probability of y being less than or equal to a.
_____ 7.	μ	g.	The probability of y being greater than or equal to a.
_____ 8.	$P(y \leqq a)$	h.	$1 - p$
_____ 9.	$P(y)$	i.	the sample mean
_____ 10.	$P(y \geqq a)$	j.	$\frac{n!}{y!\,(n - y)!} p^y q^{n-y}$

Chapter 6 STATISTICAL INFERENCE AND THE DISTRIBUTION OF THE STATISTIC

PRACTICE QUIZ

Questions

1. Given the following probability distribution for the discrete random variable y:

y	P(y)
90	.038
100	.129
110	.127
120	.244
130	.176
140	.103
150	.095
160	.088

Estimate the value of the mean.

a) 105 d) 0.244

b) 125 e) 1.000.

c) 145

2. An experiment is conducted to determine whether an untrained monkey will respond to an unknown visual stimulus. A monkey is subjected to the stimulus in four consecutive trials; after each trial the monkey is rewarded with food if he responded to the stimulus. Does the number of correct responses have a binomial distribution?

a) Yes, because at each trial there are two possible results: "response" and "no response".

b) Yes, because the probability of "response" is the same at each trial.

c) No, because the four trials lead to four possible results.

d) No, because the food reward is a reinforcement and makes the probability of responding to the stimulus different on the different trials.

e) both a) and b).

f) both c) and d).

3. A fruit basket contains three apples, two oranges, and one pear. A person picks a piece of fruit at random and randomly picks another piece of fruit after the first one is returned to the basket. An observer records for every piece picked whether it was an orange (success) or not (failure). Given that the number of oranges has a binomial distribution, what are the values of p and n?

a) $p = 1/3$; $n = 6$

b) $p = 1/2$; $n = 6$

c) $p = 1/3$; $n = 2$

d) $p = 1/2$; $n = 2$

e) The number of oranges does not have a binomial distribution.

4. A bag contains a large number of red and green marbles. 60% of the marbles are red and 40% are green. You draw 20 marbles at random from the bag and receive a prize if you draw exactly 12 red marbles. What is the probability that you will win a prize?

a) .000

b) .035

c) .120

d) .180

e) .595.

5. A multiple-choice test has 15 questions. Each question has 5 alternatives, one of which is correct. If a student guesses on every question, what is the probability of getting five or more questions correct?

a) .061 d) .164

b) .103 e) .939.

c) .146

6. A biased coin is tossed eight times. The probability of a tail is 0.4. What is the probability of obtaining exactly 3 tails?

a) .017 d) .219

b) .124 e) .279.

c) .150

7. A biased coin is tossed 16 times. The probability of a tail is 0.7. What is the probability of obtaining at least 13 tails?

a) .099 d) .245

b) .146 e) .901.

c) .219

8. Given the following probability distribution for the random variable y:

y	2	4	5	6
P(y)	0.3	0.3	0.2	0.2

Find μ, the mean of y, and σ^2, the variance of y.

a) $\mu = 4.0$; $\sigma^2 = 1.2$

b) $\mu = 4.2$; $\sigma^2 = 1.2$

c) $\mu = 4.0$; $\sigma^2 = 2.2$

d) $\mu = 4.2$; $\sigma^2 = 2.2$

e) $\mu = 4.0$; $\sigma^2 = 9.0$.

9. The probability distribution for the discrete random variable y is given by the following table:

y	$P(y)$
16	.054
21	.061
24	.083
29	.186
31	.292
35	.142
39	.105
45	.077
	1.000

Estimate the value of the standard deviation.

a) 2

b) 5

c) 15

d) 20

e) 30.

10. Consider the following table:

y	2	5	7	12
P(y)	-0.1	+0.5	+0.4	+0.2

Is this a valid probability distribution of a random variable *y*?

a) Yes, because the sum of the probabilities for all values of *y* is equal to 1.

b) Yes, because the distribution is approximately symmetric.

c) No, because the sum of *y*/*P*(*y*) is not equal to 1.

d) No, because the values of *y* have to be consecutive to make the probability distribution complete.

e) No, because negative probabilities are not possible.

Analysis

Correct Answer	Explanation of Wrong Answers
1b	a c d e see Objective 20.
2d	a b c e f see Objective 14.
3c	a *p* is correct; *n* is the number of trials, not the size of the population;
	b in a binomial distribution, *p* and *q* can take on any positive value as long as *p* + *q* = 1.0; for *n*, see answer a;
	d for *p*, see answer b; *n* is correct;
	e see Objective 14.
4d	a is *P*(0) rather than *P*(12) under *p*;
	b is *P*(12) under *q* rather than *p*;
	c is *P*(12) under *p* = .5 rather than *p* = .6;
	e is the probability of at least 12 marbles.

Correct Answer		Explanation of Wrong Answers
5d	a	is $P(y > 5)$ rather than $P(y \geqq 5)$;
	b	is $P(y = 5)$ rather than $P(y \geqq 5)$;
	c	is $P(y = 5) + P(y = 6)$;
	e	is $P(y \leqq 5)$ rather than $P(y \geqq 5)$;
6e	a	is P(0) rather than P(3) under p;
	b	is P(3) under q rather than p;
	c	you have to use the binomial table; you probably computed $(3/8) \times 0.4$;
	d	is P(3) under $p = .5$ rather than $p = .4$.
7d	a	is the probability of more than 13 tails (Study Guide Tip A);
	b	is the probability of exactly 13 tails (Study Guide Tip A);
	c	is the probability of 13 and 14 tails (Study Guide Tip A);
	e	is the probability of at most 13 tails (Study Guide Tip A).
8c	a	μ is correct; you failed to square the deviations $(y - \mu)$ in computing $\sigma^2 = \Sigma(y - \mu)^2P(y)$.
	b	you computed μ as if it were the mean of the set of scores y, rather than the mean of the random variable y with its associated probabilities; for σ^2, see answer a.
	d	for μ, see answer b; your computation of σ^2 is correct except for using the wrong value of μ;
	e	μ is correct; you failed to multiply the square of the deviations by $P(y)$ in computing $\sigma^2 = \Sigma(y - \mu)^2P(y)$.

Correct Answer	Explanation of Wrong Answers
9b	a c d e see Objective 20.
10e	a b c d see Objective 21.

Symbol Exercise Solutions

1e, 2h, 3a, 4i, 5c, 6b, 7d, 8f, 9j, 10g.

CHAPTER 7. CONTINUOUS DISTRIBUTIONS

OBJECTIVES

Chapter 7 discusses the properties of continuous distributions in order to introduce the most important continuous distribution in statistics, the normal distribution. You should be familiar with

II the properties of continuous distributions in general;

III the normal distribution;

IV the central limit theorem which spells out the importance of the normal distribution for statistical inference.

Basic to the understanding of the normal distribution are standard scores, which should be studied first (general objective I). You should be familiar with

I standard scores, which can be computed for any distribution, including the normal.

More specifically, after reading the text you should be able to

I K (1) define the terms standard score, z-score, change of scale, change of origin (Text 7.1);

C (2) distinguish between a percentile and a standard score (Text 7.1);

(3) describe the effects of change of scale and change of origin on the shape, mean, standard deviation, and variance of a distribution (Text 7.1; Study Guide Tip A);

(4) understand the advantages of converting a distribution to standard scores (Text 7.1):

a) uniformity of mean (mean = 0) and standard deviation (s.d. = 1);

b) no change in standard scores after changes of scale or origin in the original data values;

(5) interpret a standard score conversion as a change of scale and origin (Text 7.1);

A (6) convert a score $\underline{y}$ to a standard score $\underline{z}$ given the mean and standard deviation of the distribution of $\underline{y}$ (Text 7.1; Study Guide Tip B);

(7) convert a standard score $\underline{z}$ to a score $\underline{y}$, given the mean and standard deviation of the distribution of $\underline{y}$ (Text 7.1; Study Guide Tip B);

II K (8) define the terms <u>relative probability</u>, <u>probability density</u> (Text 7.2);

C (9) distinguish between continuous and discrete distributions (Text 7.2);

(10) understand the meaning of probability in a continuous distribution (Text 7.2):

a) the probability of individual values of the distribution is zero;

b) the probability of intervals is represented by the area of the distribution above the interval;

c) the total probability = the total area of the distribution = 1.0;

d) the ordinate of the distribution represents probability density;

(11) interpret the percentile of a score of a continuous distribution as the proportion of the area below the score to the total area (as for discrete distributions) (Text 7.2);

(12) interpret the mean of a continuous distribution as the balance point and the standard deviation as approximately one-half of the 68% range (as for discrete distributions) (Text 7.2);

III K (13) define the terms <u>normal distribution</u>, <u>standard normal variable</u>, <u>standard normal distribution</u>, <u>point of inflection</u>, <u>normal table</u> (Text 7.3, 7.4);

(14) list the percentage of area above the intervals ±1 s.d., ±2 s.d., ±3 s.d. for a (standard) normal distribution (Text 7.3);

(15) pronounce and interpret interval and probability notation (Text 7.4; Study Guide Tip D);

(16) list the characteristics of the "bell" shape of the normal distribution (Text 7.3);

C (17) identify the origin of the variability principle (Chapter 4) in the standard normal distribution where the area between $-1 < z < +1$ is exactly 68.26% (Text 7.3);

(18) estimate the standard deviation of a normal distribution from the distance between the mean and the points of inflection (Text 7.3);

(19) distinguish (more generally than in Chapter 4) between a forward and a backward problem (Text 7.4);

A (20) make a rough sketch of a forward or backward problem to determine which relevant area will be found in Appendix Tables A3 and A4 (Text 7.4; Study Guide Tip C);

(21) compute from the normal table the probability of intervals that

a) extend from the mean to a score z,

b) extend from a score z to infinity,

c) extend between two z-scores;

(Text 7.4);

(22) determine from the normal table the interval of standard scores with a specified probability (Text 7.4);

(23) compute the probability of an interval of non-standardized normally distributed scores and compute, in a sample of a given size, how many scores will be expected to fall in that interval (Text 7.4);

(24) determine the interval of non-standardized normally distributed scores with a specified probability (Text 7.4);

IV K (25) list the three characteristics of the distribution of the sample mean as stated by central limit theorem (Text 7.5);

C (26) interpret the distribution of the sample mean as a type of distribution of a statistic (i.e., the sample mean) (Text 7.5);

(27) describe the role of the central limit theorem in statistical inference (Text 7.5, 7.6);

(28) interpret the probability of an observed sample mean as the probability of this, or a more extreme, value in the distribution of the sample mean (Text 7.6);

A (29) compute the mean and standard deviation of the distribution of the sample mean given the population mean and standard deviation and the size of the random sample (Text 7.5, 7.6);

(30) compute the probability of a sample mean using the central limit theorem (Text 7.5, 7.6).

Figure SG 7.1 Effects of Change of Scale and Origin

		Effect on: Mean	Effect on: Standard deviation	Effect on: Shape of distribution
Change of scale	Multiply by constant A	Multiply by A	Multiply by A	No change
	Divide by constant A	Divide by A	Divide by A	No change
Change of origin	Add constant A	Add A	No change	No change
	Subtract constant A	Subtract A	No change	No change

TIPS AND REMINDERS

A. Effects of Change of Scale and Origin.

Figure SG 7.1 summarizes the effects of changes of scale and origin of a set of scores on their mean, standard deviation, and the shape of the distribution. Note that after a change of scale, the standard deviation is multiplied or divided by the same factor A as the original scores. Since the variance of a set of scores is the square of the standard deviation, the variance of the changed scores will be multiplied or divided by the square of that factor, A^2.

B. Standard Score Conversion

Negative *z*-scores present the most common source of error in standard score conversions. A score which is below the mean of its distribution has a *negative* standard score. Conversely, a negative standard score when converted to a score *y* has to be *below* the mean $\bar{y}$ (a good check for the reasonableness of your calculation).

The conversion *formulas* confirm these observations. If *y* is less than $\bar{y}$, *z* [which is $(y - \bar{y})/s$] will be negative. Conversely, if *z* is negative, *y* [which is $\bar{y} + (z \times s)$] will be less than $\bar{y}$ since $z \times s$ will be negative.

Make sure to

(a) always compute the *product* of *z* and *s* *first*, before adding it to or subtracting it from $\bar{y}$;

(b) *subtract* the product from $\bar{y}$ if *z* is a negative score.

C. Use of a Sketch in Forward and Backward Problems

To solve both forward and backward problems a rough sketch of the problem--indicating the mean of the distribution and the location of the given score, z-score, or area--is almost always necessary. (For examples, see Text Figure 7.7.)

A sketch will show you

(a) whether the area you are given or the area you are to find is in the body or in the tail of the distribution;

(b) whether a given score is above or below the mean of the distribution and thus has a positive or negative standard score;

(c) whether a given area is above or below the mean of the distribution and thus is associated with a positive or negative standard score.

D. Symbol Exercise

Match each symbol or equation in the left column with the expression that best describes it in the right column. The solutions are listed at the end of the chapter.

_____ 1. $P(0 < z < 1.34)$	a.	*y* minus $\bar{y}$, divided by *s*.
_____ 2. $\sigma_{\bar{y}}$	b.	$\bar{y}$
_____ 3. $\mu_{\bar{y}}$	c.	the probability that *z* is between 0 and 1.34.
_____ 4. z	d.	$\sigma/\sqrt{n}$
_____ 5. $\bar{y} + (z \times s)$	e.	$2 \times P(z < 1.5)$, if distribution is symmetric.
_____ 6. $P(-1.5 < z < 1.5)$	f.	"mu" sub *y*-bar

PRACTICE QUIZ

Questions

1. A set of scores has variance A. Suppose that each score of the set is divided by 4 and the variance (B) of the new scores is computed. What is the relationship between the two variances?

a) B = 16A
b) B = 4A
c) B = A
d) B = (1/4)A
e) B = (1/16)A

2. The scores on a final exam are normally distributed with a mean of 40 and standard deviation of 5. What percentage of scores lie between 25 and 55?

a) 34.13%
b) 47.72%
c) 49.87%
d) 68.26%
e) 95.44%
f) 99.74%

3. Compute the probability that a normally distributed variable has a standard score that is greater than 2.55.

a) 0.0054
b) 0.4946
c) 0.5054
d) 0.9946

4. Professor X has converted the distribution of final scores in his course to standard scores. The distribution is normal. He intends to give an "A" to the top 20% of his students. What is the cut-off standard score that guarantees an "A" in Professor X's course?

a) 0.524
b) 0.842
c) -0.524
d) -0.842
e) impossible to say without knowing the sample size.

5. Mathematics achievement test scores for 600 students were found to have a mean of 500 and standard deviation of 75. If the test scores are normally distributed, approximately how many scores would (i) fall in the interval 350 to 650? Approximately how many scores would (ii) be expected to exceed 725?

   a) (i) 95; (ii) 0

   b) (i) 95; (ii) 100

   c) (i) 300; (ii) 0

   d) (i) 570; (ii) 1

   e) (i) 570; (ii) 2.

6. Color identification of figures flashed on a screen is being studied. Thirty figures are flashed in front of each subject. It is known that the mean number of figures whose color is correctly identified is 26 and that the distribution is normally distributed. 95% of the subjects have a score between 23 and 29. What is the standard deviation of the distribution?

   a) 3.0

   b) 2.0

   c) 1.5

   d) 1.2

   e) impossible to say without further information.

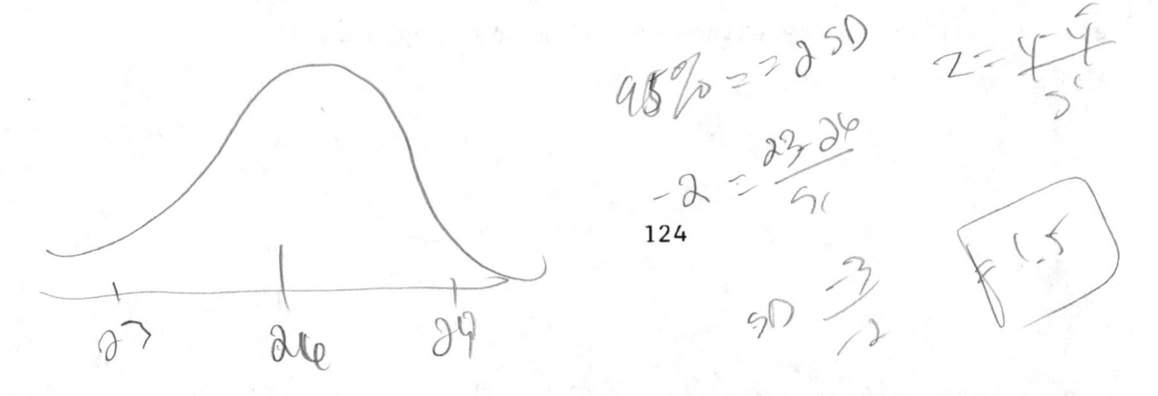

7. Suppose that the weights of adult males are normally distributed with a mean of 175 pounds and a standard deviation of 10 pounds. A certain man's weight is at the 7th percentile. What is his weight in pounds?

   a) 160 pounds
   b) 173 pounds
   c) 177 pounds
   d) 180 pounds
   e) 190 pounds.

8. Transistors manufactured by a certain process have a mean lifetime of 600 hours and a standard deviation of 45 hours. Samples of 30 transistors are taken from this group. Find the mean and standard deviation for the distribution of sample means.

   a) mean = 20; standard deviation = 1.5
   b) mean = 20; standard deviation = 8.2
   c) mean = 600; standard deviation = 1.5
   d) mean = 600; standard deviation = 8.2
   e) cannot be determined without further information.

9. Suppose that the weight of oranges is distributed with a mean $\mu = 80$ grams and a standard deviation $\sigma = 50$ grams. Samples of size n are taken and the sample mean weights are determined. 95.44% of all samples have mean weights between 70 and 90 gram. What is n, the size of the samples?

a) 3.2

b) 5

c) 10

d) 100

e) impossible to say without further information.

10. The distribution of scores obtained by high school seniors on a language aptitude test has a mean of 80 and a standard deviation of 10. This distribution is normal. Suppose that random samples of 50 scores each are taken from the original distribution of scores and the means of these samples are calculated. In what proportion of the samples would one expect the sample mean to be higher than 83?

a) .483

b) .382

c) .118

d) .021

e) .017.

Analysis

Correct Answer	Explanation of Wrong Answers
1e	a b c d see Study Guide Tip A.
2f	a b c are the percentages of scores that lie between the mean + 1 s.d. (answer a), mean + 2 s.d. (answer b), and mean + 3 s.d. (answer c). The given interval is the mean ± 3 s.d.;

Correct Answer	Explanation of Wrong Answers
	d e are the percentages of scores that lie between the mean ± 1 s.d. (answer d) and mean ± 2 s.d. (answer e).
3a	b c d you probably did <u>not</u> <u>sketch</u> the problem (see Text 7.4 and Objective 20).
4b	a c d see answer 3 b c d;
	e standard scores are independent of the sample size.
5d	a is an intermediate result, i.e. the (rounded) <u>percentages</u> of scores in the two intervals; to obtain the <u>number</u> of scores, multiply the percentages by $\underline{n}$.
	b see answer a; also (ii) is the percentage of scores that are <u>less</u> than 725;
	c you probably guessed; review Objective 23;
	e (i) is correct; (ii) is the number of scores in both tails, i.e. $\underline{P}(\underline{y} > 725) + \underline{P}(\underline{y} < 275) = 0.0026$ which, multiplied by 600, is 2.
6c	a you incorrectly divided the 95% range by two, rather than by four: the interval (mean ± 2 s.d.) which has a width of <u>4 s.d.</u> contains about 95% of a distribution;
	b d you probably guessed; review Objectives 24, 7 and 14;
	e all relevant information is provided: an interval which is symmetric around the mean, with a specified probability.

Correct Answer	Explanation of Wrong Answers
7a	b c you probably did not sketch the problem;
	d review Objectives 24 and 7;
	e see Study Guide Tip B on negative z-scores.
8d	a b c e review the central limit theorem (Text 7.5): the mean ($\mu_{\bar{y}}$) of the distribution of sample means is equal to the population mean μ, not divided by n (answers a and b); the standard deviation ($\sigma_{\bar{y}}$) of the distribution of sample meansis equal to the population standard deviation divided by the square root of n, not divided by n (answers a and c).
9d	a b c e according to the central limit theorem, $\sigma_{\bar{y}} = \sigma/\sqrt{n}$; rearranging gives $\sqrt{n} = \sigma/\sigma_{\bar{y}}$ or, by squaring, $n = (\sigma/\sigma_{\bar{y}})^2$; σ is given; to determine $\sigma_{\bar{y}}$ see Question 6.
10e	a you probably did not sketch the problem;
	b you should apply the central limit theorem: divide $(y - \mu_{\bar{y}})$ by $\sigma_{\bar{y}} = \sigma/\sqrt{n}$, not by σ;
	c see answers a and b;
	d is an intermediate result, i.e. the standard score of 83, expressed as a proportion.

Symbol Exercise Solutions

1c, 2d, 3f, 4a, 5b, 6e.

CHAPTER 8. RELATIONSHIPS BETWEEN TWO VARIABLES

OBJECTIVES

Chapter 8 discusses several aspects of the study of relationships between variables. You should be familiar with

I basic terms in the study of relationships;

two experimental designs that allow you to determine the relationship between two variables:

II two-group studies, and

IV correlational studies;

measures of relationship computed in these two types of studies:

III the standardized effect size, and

V the Pearson correlation.

More specifically, after reading the text you should be able to

I K (1) define the terms <u>statistics</u>, <u>naturally-occurring groups</u>, <u>experimentally-defined groups</u>, <u>naturally-occurring variable</u>, <u>experimentally-defined variable</u>, <u>independent variable</u>, <u>dependent variable</u> (Text 8.1, 8.2);

C (2) distinguish between naturally-occurring variables and experimentally-defined variables (Text 8.2);

(3) distinguish between studies of naturally-occurring groups (surveys) and studies of experimentally-defined groups (experiments) (Text 8.2);

(4) distinguish between a relationship observed in a sample and a relationship in a population (Text 8.1, 8.2);

II K (5) define the terms bivariate scatterplot, relationship or association between two variables (Text 8.2);

A (6) compare the results of a two-group study graphically by

a) bivariate scatterplots,

b) frequency polygons,

c) modified scatterplots (Text 8.2);

III K (7) define the terms effect size, standardized effect size, pooled standard deviation (Text 8.5);

C (8) interpret the standardized effect size as the effect size expressed in terms of the standard deviation (Text 8.5);

(9) explain the effects of a change of scale or origin of the dependent variable on the standardized effect size (Text 8.5; Study Guide Tip A);

A (10) compute the pooled standard deviation for two groups with

a) equal sample sizes (Text 8.5);

b) unequal sample sizes (Text 8.5);

(11) check a computed value of $\underline{s}'_p$ for its reasonableness: the value of $\underline{s}'_p$ must lie between the values of $\underline{s}'_1$ and $\underline{s}'_2$ (Text 8.5);

(12) compute the standardized effect size for a two-group data set (Text 8.5);

(13) estimate visually the standardized effect size for a two-group data set from the amount of overlap of the two distributions (Text 8.5);

IV K (14) define the terms correlational study, regression, regression function, regression curve (Text 8.3, 8.4);

C (15) distinguish between the regression of $\underline{x}$ on $\underline{y}$ and the regression of $\underline{y}$ on $\underline{x}$ (Text 8.4);

A (16) decide which of two continuous variables in a study to designate as the independent variable (Text 8.3);

(17) determine visually the relationship between two variables from the orientation of the regression function (Text 8.4);

(18) reduce a correlational study to a two-group study (Text 8.4);

(19) plot the regression function for a set of data with

a) a discrete independent variable;

b) a continuous independent variable (Text 8.4);

V K (20) define the terms best-fitting straight line, linear regression function, regression line, correlation, Pearson correlation, product-moment correlation (Text 8.6);

C (21) interpret the Pearson correlation as a measure of linear relationship (Text 8.6);

(22) explain the effects of a change of scale or origin of one or both variables on the Pearson correlation (Text 8.6, Study Guide Tip C);

(23) interpret a positive, negative, or zero correlation in terms of the strength of relationship between the independent variable and the dependent variable (Text 8.6);

(24) distinguish between the computational and definitional formulas of $\underline{r}$ and their respective advantages (Text 8.6);

A (25) estimate the value of $\underline{r}$ from the scatter of the sample points about the linear regression function and from the orientation of the slope of the linear regression function (Text 8.6; Study Guide Tip B);

(26) compute the Pearson correlation from original data using either the definitional or computational formula (Text 8.6);

(27) check the reasonableness of a computed value of $\underline{r}$: $\underline{r}$ has to have a value between -1 and +1 (Text 8.6).

TIPS AND REMINDERS

A. Effects of Change of Scale or Origin on Standardized Effect Size

While stated without proof in the text, it can be proven algebraically that the standardized effect size is unaffected by changes of scale or origin of the dependent variable. The standardized effect size (SES) is defined as

$$\text{SES} = \frac{\bar{y}_2 - \bar{y}_1}{s'_p}, \qquad \text{where } s'_p = \sqrt{\frac{s'^2_1 + s'^2_2}{2}}$$

Change of scale (for effects on mean and standard deviation see Text Section 7.1):

new sample means: $\bar{y}_3 = c\bar{y}_1$

$\bar{y}_4 = c\bar{y}_2$

new standard deviations: $s'_3 = cs'_1$

$s'_4 = cs'_2$

$$\text{new } s'_p = \sqrt{\frac{s'^2_3 + s'^2_4}{2}} = \sqrt{\frac{c^2 s'^2_1 + c^2 s'^2_2}{2}} = \sqrt{\frac{c^2(s'^2_1 + s'^2_2)}{2}}$$

$$= c\sqrt{\frac{s'^2_1 + s'^2_2}{2}} = c \times \text{old } s'_p$$

$$\text{new SES} = \frac{\bar{y}_4 - \bar{y}_3}{\text{new } s'_p} = \frac{c\bar{y}_2 - c\bar{y}_1}{c \times \text{old } s'_p} = \frac{c(\bar{y}_2 - \bar{y}_1)}{c \times \text{old } s'_p}$$

$$= \frac{\bar{y}_2 - \bar{y}_1}{\text{old } s'_p} = \text{old SES.}$$

Change of origin (for effects on mean and standard deviation see Text Section 7.1):

new sample means: $\bar{y}_3 = \bar{y}_1 + c$

$\bar{y}_4 = \bar{y}_2 + c$

new standard deviations: $s'_3 = s'_1$

$s'_4 = s'_2$

new s'_p = old s'_p

$$\text{new SES} = \frac{\bar{y}_4 - \bar{y}_3}{\text{new } s'_p} = \frac{(\bar{y}_2 + c) - (\bar{y}_1 + c)}{\text{old } s'_p} = \frac{\bar{y}_2 - \bar{y}_1 + c - c}{\text{old } s'_p}$$

$$= \frac{\bar{y}_2 - \bar{y}_1}{\text{old } s'_p} = \text{old SES.}$$

B. Estimating a Correlation

The value of the Pearson correlation, r, of a set of data can be estimated from the scatterplot of the data on the basis of two properties:

1) the slope of the linear regression line;

2) the scatter of the data points about the linear regression line.

The slope alone is a misleading indicator of the size of r, since it varies with the standard deviations of the x and y variables used. For example, the slope can vary for two data sets which have the same correlation (see Figure SG 8.1). Thus, when estimating a correlation from a scatterplot, the slope of the best-fitting straight line should be used only as an indicator of the sign of the correlation, i.e., positive or negative. Use the degree of scatter about the line in your estimate of the absolute value of the correlation, i.e., the larger the scatter the smaller the absolute value of r.

Figure SG 8.1 Different Slope--Same Correlation

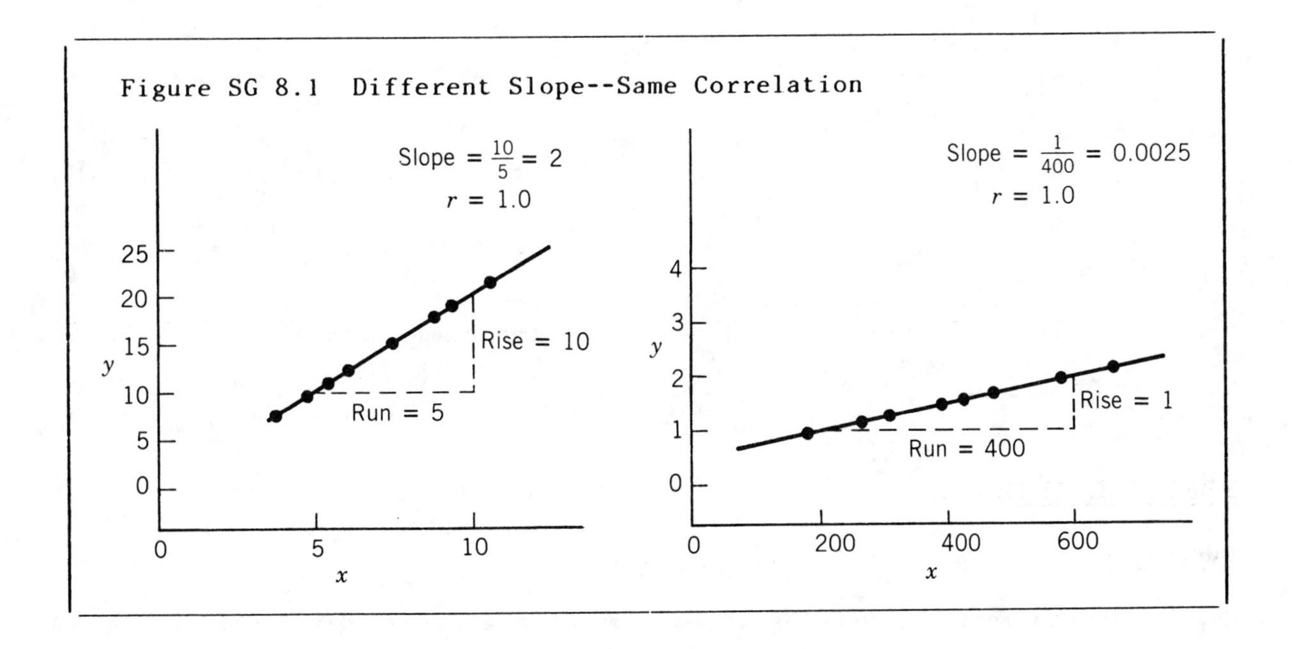

C. Symbol Exercise

Match each symbol or equation in the left column with the expression that best describes it in the right column. The solutions are listed at the end of the chapter.

_____ 1. s'_p — a. the standardized effect size

_____ 2. $\dfrac{\bar{y}_2 - \bar{y}_1}{s'_p}$ — b. the definitional formula for *r*

_____ 3. $\dfrac{\Sigma xy - (\Sigma x)(\Sigma y)/n}{n\, s'_x\, s'_y}$ — c. s'_p for two groups with unequal sample size

_____ 4. $\sqrt{\dfrac{n_1 s_1'^2 + n_2 s_2'^2}{n_1 + n_2}}$ — d. "s sub p prime", the pooled standard deviation

_____ 5. $\dfrac{\Sigma(x - \bar{x})(y - \bar{y})}{n\, s'_x\, s'_y}$ — e. the computational formula for *r*.

PRACTICE QUIZ

Questions

1. A developmental psychologist has computed a standardized effect size of 0.570 for the difference in IQ points between the twins in his sample of 22 identical twins. He later discovers that he made an error in scoring the IQ tests and has to add 2 points to every IQ score. What is the new standardized effect size?

a) 0.570
b) 0.590
c) 0.770
d) 0.855
e) 2.570.

2. Compute the pooled standard deviation for two distributions of scores with sample sizes $\underline{n}_1 = 9$, $\underline{n}_2 = 13$, means $\bar{\underline{y}}_1 = 15.9$ and $\bar{\underline{y}}_2 = 32.3$, and variances $\underline{s}_1'^2 = 4.9$ and $\underline{s}_2'^2 = 5.6$.

a) 2.29
b) 2.31
c) 5.25
d) 5.31
e) 5.32
f) 28.35.

3. In a kindergarten class of 19 boys and 19 girls, the girls' heights are distributed with a mean of 104.7 cm and a standard deviation of 5.8 cm, and the boys' heights are distributed with a mean of 109.3 cm and a standard deviation of 4.3 cm. Compute the standardized effect size of the relationship between height and sex.

a) 0.176
b) 0.901
c) 2.777
d) 3.927
e) 4.600
f) 5.105.

4. The regression curves in Figure SG 8.2 represent the regression of height (in cm) on weight (in pounds) and the regression of weight on height, for 15-year-old boys. If a randomly selected boy has a height of 135 cm, what is the best prediction of his weight?

a) 112

b) 120

c) 167

d) 180

e) impossible to say without further information.

Figure SG 8.2 Practice Quiz, Question 4

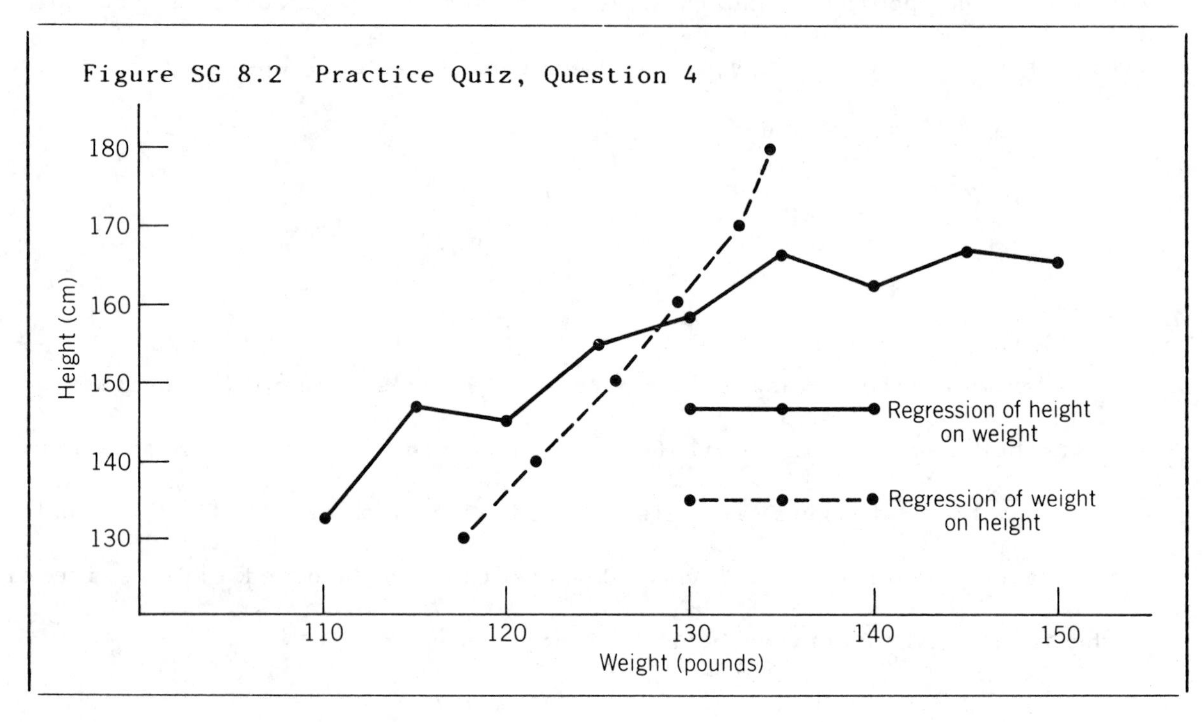

5. The correlation between the midterm and final exam grades in an introductory statistics course is 0.59. If 5 points are added to each final exam grade, the new r will be:

a) 0.71

b) 0.64

c) 0.59

d) 0.54

e) 0.12.

6. From the scatterplot given in Figure SG 8.3, which value of r gives the best estimate of the degree of correlation between x and y?

a) -0.9
b) -0.4
c) 0.0
d) +0.4
e) +0.9.

Figure SG 8.3 Practice Quiz, Question 6

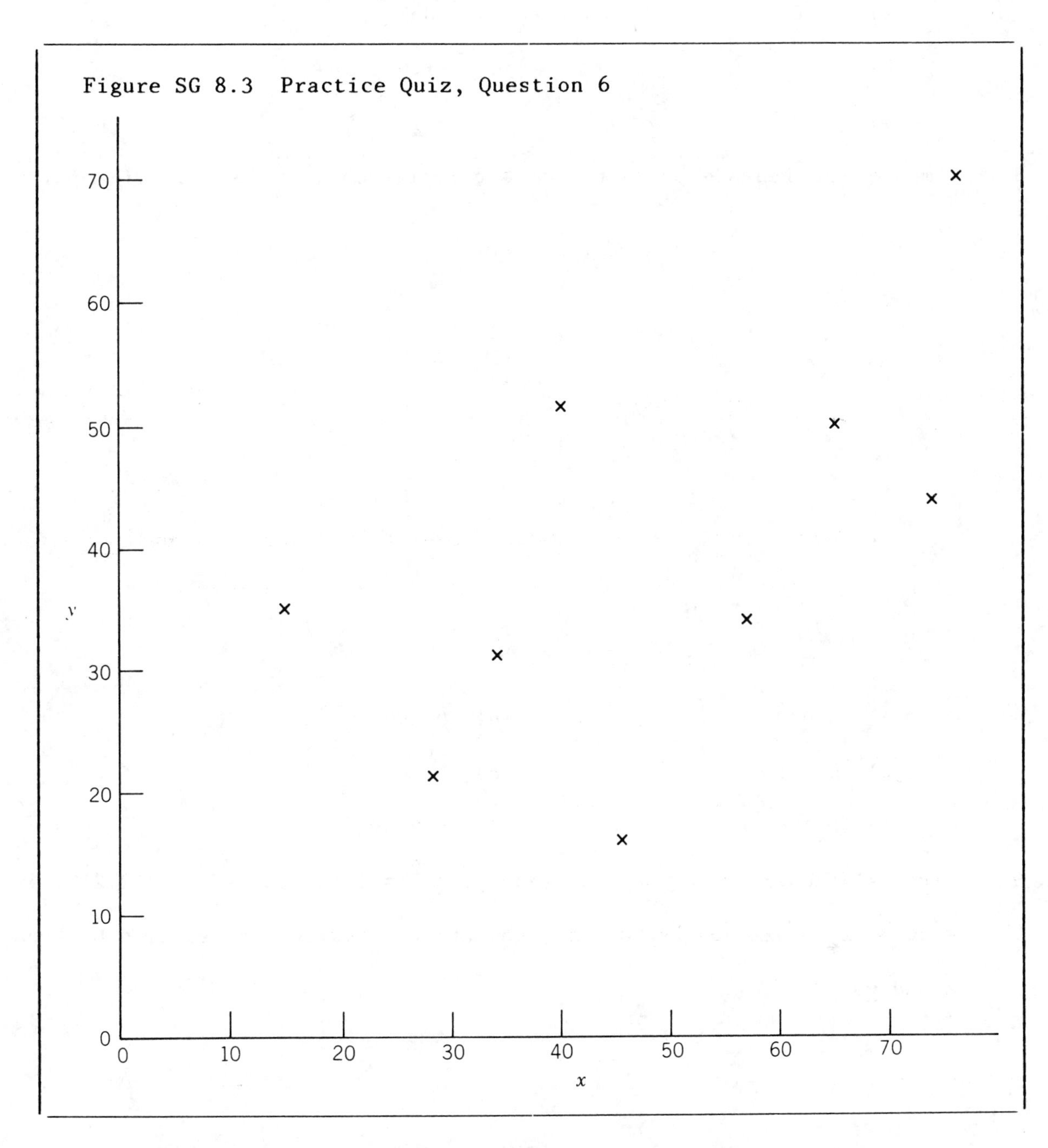

7. If performance on the Graduate Record Exam (GRE) has no influence on success in graduate school, the correlation between the GRE and success in graduate school will be:

a) +1.0 d) -0.5

b) +0.5 e) -1.0.

c) 0.0

8. Compute the Pearson product-moment correlation for the following pairs of scores:

x	y
1	1
2	7
3	1
4	4
5	7
6	1
7	7

a) 0.000 d) 0.309

b) 0.056 e) 0.707

c) 0.154 f) 0.926.

9. Given sixteen pairs of scores x and y, $s'^2_x = 1.369$ and $s'^2_y = 102.212$, $\Sigma x = 46.3$, $\Sigma y = 422$ and $\Sigma xy = 1359$, compute the Pearson correlation between x and y.

a) 0.062
b) 0.167
c) 0.728
d) impossible to say without further information.

10. Which of the correlations listed below represents the highest possible degree of relationship between two variables?

a) +0.03
b) +0.20
c) -0.22
d) +1.05
e) -1.07.

Analysis

Correct Answer	Explanation of Wrong Answers
1a	b c d e this is a change of origin; see Objective 9.
2b	a you should have used the formula of s'_p for unequal sample sizes;
	d this is s'^2_p, the pooled variance;
	c see answers a and d;
	e you probably squared the variances when computing s'_p; the given variances (4.9 and 5.6) are already the squares of standard deviations;
	f see answers d and e.
3b	a you should have divided the difference of the means by s'_p (not by s'^2_p);

Correct Answer	Explanation of Wrong Answers
	c d you probably made an error in computing s'_p, i.e., dividing $s_1^2 + s_2^2$ by n, the sample size of each group (answer c) or by $2n$ (answer d) instead of by 2;
	e f are intermediate results:
	e is the effect size,
	f is s'_p; the standardized effect size is the effect size divided by s'_p.
4b	a you used the wrong regression curve; see Objective 15;
	d you probably confused the two axes, i.e., located the given height on the weight scale;
	c see answers a and d;
	e the regression curve of weight on height provides all relevant information.
5c	a b d e this is a change of origin; see Objective 22.
6d	a b c e see Objective 25;
	a b you are wrong about the direction of the correlation;
	c the best-fitting straight line must be horizontal in order for $r = 0.0$;
	e you underestimate the scatter of the points about the best-fitting straight line.
7c	a b d e see Objective 23.

Correct Answer	Explanation of Wrong Answers
8d	a c f you probably used the definitional formula of r incorrectly:
	a you computed $\Sigma(x - \bar{x})\Sigma(y - \bar{y})$ as the numerator, which is zero for all correlations, instead of adding the product of every deviation of x from $\bar{x}$ and every deviation of y from $\bar{y}$, i.e., $\Sigma(x - \bar{x})(y - \bar{y})$;
	c you incorrectly used the number of observations as n instead of the number of pairs of scores;
	f you ignored the signs of the deviations of x from $\bar{x}$ and the deviations of y from $\bar{y}$, treating them, incorrectly, as absolute values;
	b you incorrectly used the number of observations as n instead of the number of pairs of scores; in the computational formula for r.
	e you used the variances instead of the standard deviations.
9c	a you used the variances of x and y instead of their standard deviations;
	b see answer a; you also used the number of observations as n instead of the number of pairs of scores;
	d all relevant information is provided: Σx, Σy, Σxy, s_x, s_y, n.
10c	a b d e see Objectives 27 and 23.

Symbol Exercise Solutions

1d, 2a, 3e, 4c, 5b.

REVIEW CHAPTER B

OBJECTIVES

Chapter B summarizes Chapters 6 and 7 by discussing the common structure of statistical inference (Chapter 6) which applies equally to inferences about a proportion (Chapter 6) and inferences about a mean (Chapter 7). You should review Chapters 6 and 7 to be familiar with these topics.

More specifically, after reading Chapter B you should be able to

K (1) list the common steps of statistical inference (Text B.2);

C (2) distinguish between an analysis of a mean (for continuous variables) and an analysis of a proportion (for two-valued variables) (Text B.1);

(3) contrast the differences in detail between making an inference about a proportion or an inference about a mean (Text B.2);

(4) distinguish between the use of the symbol "$\underline{y}$" in the analysis of a mean and in the analysis of a proportion (Text B.2);

(5) explain the relationship between the possible values of μ and $\underline{p}$ in a population (Text B.1);

A (6) decide whether to analyze a given population by making an inference about its mean or about its proportion of successes (Text B.1; Study Guide Tip B);

TIPS AND REMINDERS

A. Symbol Exercise--Review of Chapters 2 to 8

Match each symbol or equation in the left column with the expression that best describes it in the right column. The solutions are listed at the end of the chapter.

____ 1. $\bar{y}$	a.	$n \times \frac{\text{percentile}}{100}$
____ 2. s'^2	b.	$\frac{\text{frequency}}{n}$
____ 3. p	c.	sigma, the standard deviation of a random variable
____ 4. $\frac{\bar{y}_2 - \bar{y}_1}{s'_p}$	d.	$\sigma/\sqrt{n}$
____ 5. μ	e.	$\frac{\Sigma y}{n}$
____ 6. Σ	f.	the probability of y being greater than or equal to a
____ 7. σ	g.	$\frac{\Sigma(y - \bar{y})^2}{n}$
____ 8. r	h.	the standardized effect size

____ 9.	cumulative frequency below midpoint	i.	the probability of success
____ 10.	$\sigma_{\bar{y}}$	j.	$\frac{n!}{y!\,(n-y)!}\,p^y q^{n-y}$
____ 11.	$\mu_{\bar{y}}$	k.	sigma, "the sum of"
____ 12.	relative frequency	l.	$\frac{y - \text{mean}}{\text{s.d.}}$
____ 13.	z	m.	$\Sigma yP(y)$
____ 14.	P(y)	n.	"mu sub y-bar", the mean of the distribution of the statistic $\bar{y}$
____ 15.	$P(y \geqq a)$	o.	Pearson correlation

B. Test of a Mean vs. Test of a Proportion

Some populations, like the example in Text Chapter B--the population of the length of time psychiatric patients on a certain psychoactive drug remain in the community after discharge from the hospital--can be analyzed by either making an inference about their mean or about the proportion of successes, where a success can be defined in many ways.

The decision on which method to use depends on several factors. Depending on how much weight is attributed to each factor, different decisions will be made:

a) Practicality of data collecting. For the example in the text, it is considerably easier to keep track of the psychiatric patients for two years than for ten years or even longer. Time and financial resources have to be considered.

b) Purpose of the analysis. For a pilot study comparing the effectiveness of two different psychoactive drugs, the proportion of patients in the community after two years is an adequate test statistic. To make a definitive statement about the effectiveness of one particular psychoactive drug in keeping patients in the community, an inference about the mean length of stay in the community is necessary.

c) Fair and efficient use of collected data. For the example in the text, if the patients were followed for only two years, the proportion of successes, i.e., patients that remain in the community, should be used. It would be misleading to use the mean length of stay in the community as the test statistic, since the mean stay could only be computed for those patients that have returned to the hospital, excluding all the "successful" patients that are still in the community. If the patients were followed for ten years, with some still in the community after this time, the same problem exists, i.e., unfairly excluding the "successful" patients in the computation of the mean. However, this disadvantage is probably far outweighed by the loss of information that results when the data collected over ten years are just divided into the two groups of success or failure at the two-year point or five-year point. It may appear as a solution to wait until all patients in the study have returned to the hospital and then to compute the mean, but chances are that a few patients or even the investigator would die before that time.

Symbol Exercise Solutions

1e, 2g, 3i, 4h, 5m, 6k, 7c, 8o, 9a, 10d, 11n, 12b, 13ℓ, 14j, 15f.

CHAPTER 9. HYPOTHESIS TESTING: THE BASICS

OBJECTIVES

Chapter 9 discusses in detail the techniques of hypothesis testing, some of which were already introduced in Chapters 6, 7, and B. You should be familiar with

I the basic steps of hypothesis testing and their rationale;

two particular applications of hypothesis testing:

II the binomial test, and

III the z-test of one mean;

IV the normal approximation to the binomial distribution as a method of determining binomial probabilities.

More specifically, after reading the text you should be able to

I K (1) define the terms *specific hypothesis*, *simple hypothesis*, *acceptance region for H_o*, *acceptance region for H_1*, *critical value*, *Type I error*, *Type II error*, *probability of a Type I error* (α), *probability of a Type II error* (β) (Text 9.1);

C (2) distinguish between a Type I and a Type II error (Text 9.1);

(3) explain the relationship between the probability of a Type I error (α) and the probability of a Type II error (β) (Text 9.1; Study Guide Tip A);

(4) explain the basis of the choice of the critical value (Text 9.1);

(5) explain how to locate the acceptance regions, based on the relative locations of the H_o and H_1 distributions. (Text 9.1);

II K (6) define the terms binomial test, cumulative probability, test statistic (Text 9.1, 9.2, 9.3);

(7) use inequality symbols to express cumulative probabilities (Text 9.2);

(8) state which test statistic is used in the binomial test (Text 9.1);

C (9) explain why the statistic y has a different distribution under the two hypotheses (Text 9.1);

(10) interpret the critical value as the "real limit" between the acceptance region of H_o and the acceptance region of H_1 in the binomial test (Text 9.1);

A (11) draw a sketch of the distributions of y under H_o and H_1, indicating the means (= np), in order to decide in which tail of the H_o distribution to locate α (Text 9.1; Study Guide Tip A);

(12) compute cumulative tail probabilities using

a) the binomial table

b) the cumulative binomial table (for $\underline{n}$ = 5, 10, 15, 20, 25) (Text 9.2);

(13) compute individual or interval probabilities using

a) the binomial table

b) the cumulative binomial table (for $\underline{n}$ = 5, 10, 15, 20, 25) (Text 9.2);

(14) determine the acceptance regions for $\underline{H}_o$ and $\underline{H}_1$:

a) find the critical value of $\underline{y}$ which fulfills the given restriction on α (backward problem) (Text 9.1);

b) decide on which side of the critical value to locate the acceptance regions of $\underline{H}_o$ and $\underline{H}_1$ (Text 9.1);

(15) make a decision about $\underline{H}_o$ or $\underline{H}_1$ based on the acceptance region in which the observed $\underline{y}$ falls (Text 9.2);

(16) compute β for a given critical value, given $\underline{n}$ and the distribution of the statistic $\underline{y}$ under $\underline{H}_1$ (forward problem) (Text 9.1);

III K (17) define the term $\underline{z}$-test (Text 9.3);

(18) state which test statistic(s) is/are used in the $\underline{z}$-test of one mean (Text 9.3);

C (19) distinguish between the two test statistics $\bar{\underline{y}}_{obs}$ and $\underline{z}_{obs}$ (Text 9.3);

A (20) draw a sketch of the distributions of $\bar{y}$ under H_o and H_1 (using the central limit theorem) to decide on the locations of α and β and the two acceptance regions (Text 9.3; Study Guide Tip A);

(21) compute the critical value of $\bar{y}$:

a) find its value in standard scores, z_{crit} (Text 9.3);

b) convert z_{crit} to a score in a distribution with mean = μ_o and standard deviation = $\sigma_{\bar{y}}$ (Text 7.1);

(22) locate the acceptance regions of H_o and H_1 (Text 9.3);

(23) convert the observed sample mean into a standard score of the H_o distribution (Text 7.1);

(24) express the acceptance regions of H_o and H_1 in terms of

a) $\bar{y}_{crit}$

b) z_{crit} (Text 9.3);

(25) make a decision about H_o or H_1 based on the acceptance region in which the observed $\bar{y}$ falls (Text 9.3);

(26) compute β for a given critical value given the distribution of $\bar{y}$ under H_1 (Text 9.3);

IV K (27) state the accuracy rules for the normal approximation to the binomial distribution (Text 9.4);

(28) decide when to use the normal approximation to determine binomial probabilities (Text 9.4);

C (29) understand how the following topics relate to the technique of approximating the binomial distribution by the normal distribution: real limits, use of the normal table, the principles of reasonableness and usefulness, and standard score conversion (Text 9.4);

A (30) carry out the normal approximation to obtain individual or cumulative binomial probabilities (Text 9.4).

TIPS AND REMINDERS

A. Relationship between α and β

The probability of making a wrong decision when H_o is true, α, is an area of the distribution of the test statistic when H_o is true. The probability of making a wrong decision when H_1 is true, β, is an area of the distribution of the test statistic when H_1 is true. Thus, α and β are never represented by areas of the same distribution. They are, however, related by the critical value which, all other factors constant, depends on the choice of α. The critical value in turn determines the value of β.

Whether α and the critical value are in the upper or the lower tail of the H_o distribution, β is the area of the H_1 distribution located on the side of the critical value opposite to α (see Text Figures 9.3, 9.4, 9.6, 9.7b).

Many mistakes in hypothesis testing can be avoided by drawing a careful sketch of the distributions of the test statistic under the two hypotheses. This sketch should indicate α, the critical value, and β. It is important that both distributions use the same scale so that visually the distribution with a higher mean will be to the right of the other distribution. Also, you should compute and indicate the standard deviations of the two distributions. Decide in which tail to place α and compute the critical value. Then draw a vertical line through the critical value that will divide both distributions into two parts. The area in the H_1 distribution which is on the other side of the critical value (that is, opposite α in the H_o distribution) is β.

If your sketch is reasonably accurate, the size of the area representing β relative to the whole H_1 distribution is a good estimate of β against which you can check your computed value for its reasonableness.

B. Cumulative Probability Table

Figure SG 9.1 illustrates how the values of the cumulative binomial table (A2) are derived from the values of the simple binomial table (A1). For $n = 10$ and $p = .5$, the histogram represents the probabilities of each value of y (Table A1). The polygon shows the cumulation of these individual probabilities into cumulative probabilities (Table A2) for the interval between 0 and y. Note that the cumulative probability reaches exactly 1.000 only at $y = 10.5$.

Figure SG 9.1 Individual and Cumulative Probabilities

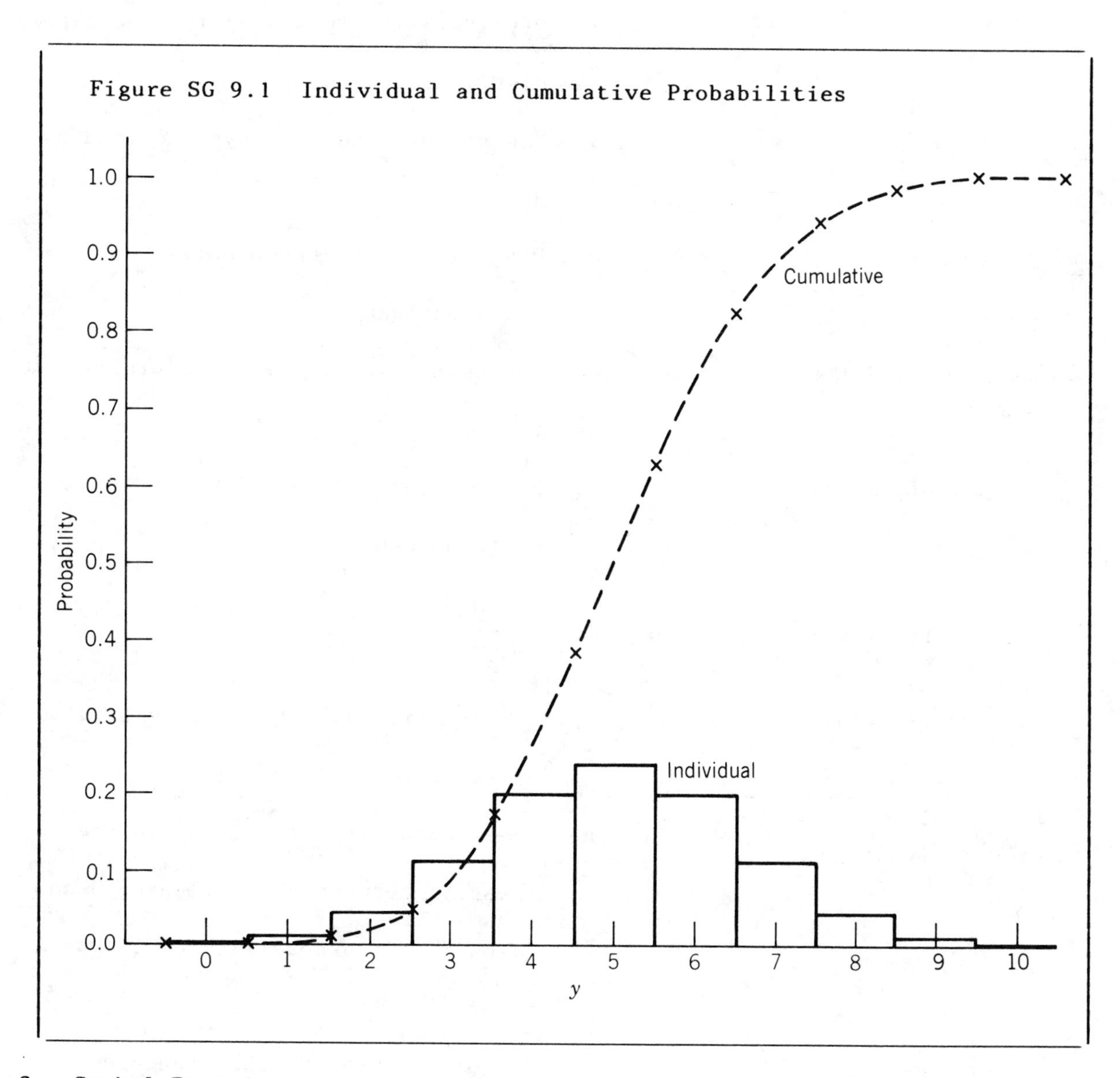

C. Symbol Exercise

Match each symbol or equation in the left column with the expression that best matches it in the right column. The solutions are listed at the end of the chapter.

Chapter 9 HYPOTHESIS TESTING: THE BASICS

_____ 1.	y	a.	$P(y < 6.5) - P(y < 5.5)$ for the binomial distribution
_____ 2.	α	b.	"mu sub 0", the mean hypothesized by H_o
_____ 3.	β	c.	$P(y < 12.5)$ for the binomial distribution
_____ 4.	$P(y \leqq 12)$	d.	the test statistic for the binomial test
_____ 5.	$P(y \geqq 13)$	e.	$P(y > 12.5)$ for the binomial distribution
_____ 6.	$P(3 \leqq y \leqq 7)$	f.	$\frac{\bar{y} - \mu_o}{\sigma_{\bar{y}}}$
_____ 7.	$P(y = 6)$	g.	"y-bar critical", the critical value of the z-test of one mean
_____ 8.	$\bar{y}_{crit}$	h.	the probability of y lying in the interval between and including 3 and 7
_____ 9.	μ_o	i.	"alpha", the probability of a Type I error
_____ 10.	$\sigma_{\bar{y}}$	j.	"sigma sub y-bar", the standard deviation of the distribution of the statistic $\bar{y}$
_____ 11.	z_{obs}	k.	"beta", the probability of a Type II error

PRACTICE QUIZ

Questions

1. The probability of a Type I error when $\beta = 0.05$ is:

    a) .025

    b) .05

    c) .975

    d) .95

    e) insufficient information to answer.

2. Prior to a federal election in Canada, a poll of a sample of fifteen people is taken to determine whether B would be preferred as Prime Minister to A. If the preferences for the two are equal, then the probability that a given person will prefer A would be 0.5 (H_o). If more people in the population prefer B, then this probability, p, that a given person will prefer A, would be 0.3 (H_1).

    What is the probability under H_1 that at least 7 people in the sample of 15 prefer A?

    a) .050 d) .696

    b) .081 e) .869

    c) .131 f) .950.

3. For $p = 0.8$ and $n = 25$, find $P(14 < y \leq 19)$.

    a) .000 d) .377

    b) .214 e) .381.

    c) .218

4. An experiment studies the effect of therapy on reducing psychotic behavior. Fifteen patients received therapy and are then rated on whether their psychotic behavior has been reduced. If the therapy has no effect then the probability of reduced psychotic behavior would be 0.5 (H_o). If the therapy has an effect then this probability, p, would be 0.6 (H_1).

   Let the test statistic, y, be the number of patients, in the sample of fifteen, whose psychotic behavior is reduced. Answer the question so that the probability of a Type I error is less than 0.10 (and the probability of a Type II error is as small as possible).

   What is the acceptance region for the hypothesis of no effect?

   a) $y < 4.5$ d) $y > 10.5$

   b) $y > 4.5$ e) $y < 11.5$

   c) $y < 10.5$ f) $y > 11.5$.

5. Refer to Question 4. The same experiment is repeated with a sample of 10 patients. The experiment determines the critical value for $\alpha \leqq .10$ as $y_{crit} = 7.5$. If seven of the ten patients show reduced psychotic behavior, what conclusion(s) do you draw?

   a) Accept H_o, because y falls in the acceptance region for H_o.

   b) Accept H_1, because y falls in the acceptance region for H_1.

   c) Accept H_o, because $y > y_{crit}$.

   d) Accept H_1, because $y < y_{crit}$.

e) Both a) and c) are proper conclusions.

f) Both b) and d) are proper conclusions.

6. Refer to Question 5. What is the value of β for this experiment ($\underline{H}_o$: $\underline{p}$ = 0.5; $\underline{H}_1$: $\underline{p}$ = 0.6; $\underline{n}$ = 10; $\alpha \leqq .10$; $\underline{y}_{crit}$ = 7.5)?

a) .012

b) .167

c) .833

d) .945

e) .988.

7. The heights of 10-year-old boys in the U.S. had a mean of 131 cm and a standard deviation of 10 cm in 1930. A random sample of 100 boys of age 10 is taken in 1980 to determine whether the population mean is still the same, i.e., 131 ($\underline{H}_o$), or has increased by 5 cm, i.e., 136 ($\underline{H}_1$). Assume that the population standard deviation does not change.
What is the critical value of the test statistic $\bar{\underline{y}}$, the mean height in a random sample of 100 boys, if α is set at 0.05?

a) 147.4

b) 137.6

c) 134.4

d) 132.6

e) 129.4

f) 114.6.

8. An experimental psychologist wants to test the intensity threshold for hearing a tone which has a frequency of 400 cycles per second. The subjects in her experiment are given a warning when to expect the tone. The question is whether such subjects have the same threshold as subjects who are not given a warning ($\underline{H}_o$: μ = 25 db) or whether they have a lower threshold than subjects who are not given a warning ($\underline{H}_1$: μ

= 20 db). Decibels (db) are a measure of the loudness of a tone. The experimenter knows that the population standard deviation is 3 db. The critical value of the test statistic, z_{obs}, is $z_{crit} = -2.326$ for $\alpha = .01$.

In a random sample of 50 subjects who were told when to expect the tone, the mean threshold is found to be 23.9 db. What should the psychologist conclude?

a) Accept H_o, because $\bar{y}$ falls in the acceptance region for H_o.

b) Accept H_1, because $\bar{y}$ falls in the acceptance region for H_1.

c) Accept H_o, because $z_{obs} > z_{crit}$.

d) Accept H_1, because $z_{obs} < z_{crit}$.

e) Both a) and c) are proper conclusions.

f) Both b) and d) are proper conclusions.

9. Given H_o: $\mu = 65$ and H_1: $\mu = 72$ with $\sigma = 3.3$, what is the acceptance region of H_o for a random sample of size 30 if the probability of a Type I error should not exceed 0.05?

a) $\bar{y} < 1.64$ d) $\bar{y} > 1.64$

b) $\bar{y} < 65.99$ e) $\bar{y} > 65.99$

c) $\bar{y} < 70.43$ f) $\bar{y} > 70.43$.

10. Compute the probability of a Type II error if the critical value of $\bar{y}$ for a random sample of size 60 is 16.485, given H_o: $\mu = 17.3$, H_1: $\mu = 15.9$ and $\sigma = 2.45$.

a) .005 d) .595

b) .032 e) .968

c) .405 f) .995.

Analysis

Correct Answer	Explanation of Wrong Answers
1e	a b c d α and β are not directly related; see Objectives 2 and 3.
2c	a you computed the probability that more than 7 people prefer A, instead of 7 or more;
	b you computed the probability that exactly 7 people prefer A;
	d is the probability that at least 7 people prefer A under H_o: $p = .5$;
	e f you probably used the cumulative binomial table incorrectly: e and f are lower cumulative probabilities; f is the probability of at most 7 people favoring A; e is the probability of less than 7 people favoring A.
3d	a you probably used the cumulative binomial table incorrectly, i.e., matched p for the lower cumulative probability with the y column for the upper cumulative probability; otherwise, your procedure is correct;

Correct Answer	Explanation of Wrong Answers
	b c e see Objective 7:
	b is $P(14 < y < 19)$,
	c is $P(14 \leqq y < 19)$,
	e is $P(14 \leqq y \leqq 19)$.
4c	b you may have put α into the wrong tail of the H_o distribution; see Objective 11. If that is not the case, you probably used the cumulative binomial table incorrectly; you should look up the upper cumulative probability;
	d is the acceptance region for H_1; see Objective 11 for the importance of a sketch;
	a see answers b and d;
	e you used $\alpha \leqq .05$ instead of .10;
	f see answers d and e;
5a	b c d e f see Objective 15 and Objective 11 for the importance of a sketch.
6c	a e you probably used the cumulative binomial table incorrectly, i.e., matched p for the lower cumulative with the y column for the upper cumulative, or vice versa; otherwise, your procedure is correct;
	b is $1.0 - \beta$; see Objective 11 for the importance of a sketch;

Correct Answer	Explanation of Wrong Answers
	d is $P(\bar{y} < \bar{y}_{crit})$ under H_o rather than H_1; this probability is $1.0 - \alpha$.
7d	a you failed to apply the central limit theorem: the standard deviation of the distribution of $\bar{y}$ under H_o and H_1 is $\sigma_{\bar{y}} = \sigma/\sqrt{n}$ not σ;
	c you used μ_1 instead of μ_o in your conversion of z_{crit} to $\bar{y}_{crit}$; see Objectives 21 and 20;
	e you put α into the wrong tail of the H_o distribution; see Objective 20 for the importance of a sketch;
	f see answers a and e;
	b see c and e.
8f	a c e see answer 7a;
	b d both b and d are proper conclusions; thus the correct answer is f.
9b	a is an intermediate result, i.e., z_{crit}; see Objective 21b;
	c see answer 7a;
	e is the acceptance region for H_1; see Objective 20;
	d see answers a and e;
	f see answers c and e.
10b	e is $1.0 - \beta$; see Objective 20 for the importance of a sketch;

Correct Answer	Explanation of Wrong Answers
	f is $\underline{P}(\bar{\underline{y}} > \bar{\underline{y}}_{crit})$ under $\underline{H}_o$ rather than $\underline{H}_1$ (see Objective 26); this probability is 1.0 - α;
	a you made two errors; you computed $\underline{P}(\bar{\underline{y}} < \bar{\underline{y}}_{crit})$ under $\underline{H}_o$, rather than $\underline{H}_1$ (see Objective 26); furthermore, the inequality should be $\bar{\underline{y}} > \bar{\underline{y}}_{crit}$; the computed probability is α; see Objective 20 for the importance of a sketch;
	c you failed to apply the central limit theorem; see answer 7a;
	d you made two errors; you computed 1.0 - β; see Objective 20 for the importance of a sketch; you also failed to apply the central limit theorem; see answer 7a;

Symbol Exercise Solutions

1d, 2i, 3k, 4c, 5e, 6h, 7a, 8g, 9b, 10j, 11f.

CHAPTER 10. HYPOTHESIS TESTING: FURTHER CONSIDERATIONS

OBJECTIVES

Chapter 10 elaborates the basic outline of hypothesis testing given in Chapter 9. You should be familiar with the notions of

I testing a null hypothesis against an alternative hypothesis;

II a significance test;

III two-sided alternative hypothesis, and

IV the p-value procedure.

Also, as another application of the theory of hypothesis testing, you should be familiar with

V the z-test of two independent means.

More specifically, after reading the text you should be able to

I K (1) define the terms composite hypothesis, null hypothesis, alternative hypothesis, acceptance region for $\underline{H}_o$, acceptance region for $\underline{H}_1$ (Text 10.1);

C (2) explain the effect of a change in the specific value of $\underline{H}_1$ on the acceptance regions for $\underline{H}_o$ and $\underline{H}_1$ (Chapter 9) and on α and β (Text 10.1);

(3) distinguish between specific hypotheses and composite hypotheses (Text 10.1);

A (4) establish the value of H_o and the direction of H_1 for a given situation (Text 10.1);

(5) make a sketch of a situation showing the H_o distribution of the test statistic and the relative location of the H_1 distribution (Text 10.1);

(6) determine the critical value of the test statistic:

a) y_{crit}, in a test of a proportion, given α, n, H_o, and H_1 (Text 10.1);

b) $\bar{y}_{crit}$ or z_{crit}, in a test of a mean, given α, σ, n, H_o, and H_1 (Text 10.1);

(7) establish the acceptance regions for H_o and H_1, respectively, based on the critical value and the direction of H_1 (Text 10.1);

II K (8) define the terms <u>significance test</u>, <u>significant result</u>, <u>non-significant result</u>, <u>rejection region</u>, <u>acceptance region</u> (Text 10.2);

(9) state the objective of a significance test (Text 10.2);

C (10) explain why H_o cannot be accepted (Text 10.2; Study Guide Tip A);

(11) distinguish between a statistically reliable (= significant) result and a result at chance level (= nonsignificant) (Text 10.2);

(12) distinguish between statistical and scientific significance (Text 10.2);

(13) state two restrictions on generalizations made from a statistically significant result (Text 10.2);

A (14) report a significant or a nonsignificant result in different ways (Text 10.2);

III K (15) define the terms _one-sided alternative hypothesis_, _two-sided alternative hypothesis_, _nondirectional hypothesis_, _one-tailed test_, _two-tailed test_ (Text 10.3);

(16) list the differences between a one-tailed test and a two-tailed test (Text 10.3; Study Guide Tip B);

(17) list the reasons that favor the use of two-tailed tests (Text 10.3);

C (18) distinguish between a one-sided alternative hypothesis and a two-sided alternative hypothesis (Text 10.3);

(19) decide whether to use a one-tailed or a two-tailed test in a given situation (Text 10.3);

A (20) compute the two critical values of the test statistic in a two-tailed test

a) $\underline{y}_{crit}$, in a test of a proportion given $\underline{H}_o$, α, and $\underline{n}$ (Text 10.3);

b) $\bar{\underline{y}}_{crit}$ or $\underline{z}_{crit}$, in a test of a mean, given α, σ, $\underline{n}$, and $\underline{H}_o$ (Text 10.3);

(21) establish the acceptance region and the rejection region in a two-tailed test of $\underline{H}_o$ (Text 10.3);

(22) decide between $\underline{H}_o$ and $\underline{H}_1$ based on the observed value of the test statistic for one- and two-tailed tests (Text 10.1, 10.3);

IV K (23) define the terms rejection region procedure, p-value procedure, p-value (Text 10.5);

(24) list the steps in testing a null hypothesis against an alternative hypothesis using the rejection region procedure (Text 10.4);

(25) list the steps in testing a null hypothesis against an alternative hypothesis using the p-value procedure (Text 10.5);

(26) list two factors that influence the size of the p-value (Text 10.5);

C (27) distinguish between a one-tailed p-value and a two-tailed p-value (Text 10.5);

(28) distinguish between the rejection region procedure and the p-value procedure for testing a null hypothesis against an alternative hypothesis (Text 10.5);

(29) explain the "dangers" in the p-value procedure (Text 10.5);

A (30) compute the p-value of the observed value of a test statistic for

a) one-tailed tests (Text 10.5);

b) two-tailed tests (Text 10.5);

(31) decide between H_o and H_1 on the basis of a comparison between the p-value of the observed result and α (Text 10.5);

V K (32) list the properties of the distribution of the test statistic $\bar{y}_1 - \bar{y}_2$ in the z-test of two independent means (Text 10.6);

A (33) establish H_o and H_1 and the value of the test statistic $\bar{y}_1 - \bar{y}_2$ for a given situation (Text 10.6);

(34) convert the value of the statistic $\bar{y}_1 - \bar{y}_2$ into a standard score given the population standard deviations, σ_1, and σ_2, and the two sample sizes (Text 10.6);

(35) carry out a significance test on the difference between two independent means with z_{obs} as test statistic using

a) the rejection region procedure (Text 10.6);

b) the p-value procedure (Text 10.6).

TIPS AND REMINDERS

A. The Probability of a Type II Error

To compute β, the probability of a Type II error, i.e., the probability of accepting H_o when the alternative hypothesis is in fact true, you have to specify an exact value for the parameter in the alternative hypothesis, H_1. (See Chapter 9 for H_1 as a specific hypothesis.) For example, in a one-tailed test of a proportion in which the parameter, p_1, specified by the alternative hypothesis is less than the null hypothesis value of the parameter, β is equal to $P(y > y_{crit})$. This probability can only be calculated if p_1 is known. Similarly, in the one-tailed test of a mean in which the parameter, μ_1, specified by the alternative hypothesis is less than the null hypothesis value, β is equal to $P(\bar{y} > \bar{y}_{crit})$. Again, this probability can only be calculated if μ_1 is known.

When the value of the parameter for the alternative hypothesis is not known, β cannot be determined.

Note that one cannot accept H_o if β cannot be determined even when the test statistic falls into the acceptance region for H_o, because the probability that a Type II error has been committed is not known. The only legitimate conclusion in this case is that one "fails to reject H_o".

B. Differences Between a One-Tailed Test and a Two-Tailed Test

One-tailed	Two-tailed
a) $\underline{H}_1$ is one-sided;	a) $\underline{H}_1$ is two-sided
b) rejection region consists of one interval of values;	b) rejection region consists of two intervals of values;
c) one critical value;	c) two critical values;
d) α = probability of one tail of the distribution of the test statistic under $\underline{H}_o$ (critical value and more extreme values);	d) α = probability of both tails of the distribution of the test statistic under $\underline{H}_o$ (critical values and more extreme values);

e) Strictly speaking, the conclusion from a one-tailed test is that the result is "significantly greater than" (or "significantly less than") the hypothesized value, and the conclusion from a two-tailed test is that the result is "significantly different from" the hypothesized value, corresponding to different forms of $\underline{H}_1$ for the two tests. However, this distinction in the reporting of the results of one- and two-tailed tests is not usually made, and the results of both tests are usually reported as "significantly greater than" (or "significantly less than") the hypothesized value.

C. Inequalities for Negative Values of z

In two-tailed z-tests, or in one-tailed z-tests with an alternative which hypothesizes lower values of the population parameter than H_o, you will encounter critical values of z that are negative. For example, for H_o: $\mu = 10$, H_1: $\mu < 10$, and $\alpha = 0.05$, we find that $z_{crit} = -1.645$.

In these cases, you should take extra care to sketch the distributions of the test statistic under H_o and H_1, indicating on the sketch the locations of α, z_{crit}, and z_{obs}. (Because H_1 is a composite hypothesis, you can only sketch its location <u>relative</u> to the H_o distribution, i.e., to the left of H_o if H_1: $\mu < \mu_o$, to the right of H_o if H_1: $\mu > \mu_o$, and to the right <u>and</u> left of H_o if H_1: $\mu \neq \mu_o$.)

A sketch will help you to establish the correct acceptance and rejection regions and to make the correct conclusion. For the above example, $z > -1.645$ is the acceptance region for H_o, and $z < -1.645$ the rejection region for H_o. A value of $z_{obs} = -1.8$ would lead to rejecting H_o, since $-1.8 < -1.645$ (even though in absolute values $|-1.8| > |-1.645|$). A value of $z_{obs} = -1.4$ would lead to a failure to reject H_o, since $-1.4 > -1.645$ (even though in absolute values $|-1.4| < |-1.645|$). Make sure you understand how these negative numbers are treated.

PRACTICE QUIZ

Questions

1. The student council claims that 70% of all students favor an increase in student fees to subsidize a new recreational area. The student newspaper sets out to test whether this proportion is less than 70%, and interviews a random sample of 25 students. What is the critical value of y, the number of students in favor of an increase, if the probability of a Type I error is not to exceed 0.01 in a one-tailed test?

   a) 10.5 d) 21.5
   b) 11.5 e) 22.5
   c) 12.5 f) 23.5.

2. Final grades in conventional statistics courses are known to be normally distributed with a mean of 69% and a standard deviation of 8%. An innovative instructor claims that a PSI course would raise the mean final grade. (Assume that the standard deviation does not change.) What is the acceptance region for H_1, expressed in terms of z, if the probability of a Type I error is not to exceed 0.05 in a one-tailed test?

   a) $z > 1.645$ d) $z < -0.126$
   b) $z < 1.645$ e) $z > -1.645$
   c) $z > 0.126$ f) $z < -1.645$.

3. In 1975, raw materials accounted for 60% of Canadian exports. In 1980 a customs official examines a random sample of 25 export shipments to see

whether the percentage of raw material shipments has changed. What are the two critical values of y, the number of raw material shipments in the sample, for this test if the probability of a Type I error is set at 0.10?

a) 5.5; 14.5
b) 6.5; 13.5
c) 10.5; 19.5
d) 11.5; 18.5
e) 14.5; 19.5.

4. The weight of North American infants at birth is normally distributed with a mean of 3.8 kg and a standard deviation of 0.3 kg. The department of health has selected a random sample of 100 pregnant women who are vegetarians to test whether a vegetarian diet during pregnancy affects the infant's weight at birth. What are the two critical values of $\bar{y}$, the mean weight at birth in a sample of 100 infants, for this study if α should not exceed 0.01?

a) -2.326; 2.326
b) -2.576; 2.576
c) 3.027; 4.573
d) 3.102; 4.498
e) 3.723; 3.877
f) 3.730; 3.870.

5. At a Midwestern university, 70% of all graduate students that are admitted finish their degree program successfully. A new admissions policy has been introduced, and a random sample of 20 graduate students is monitored to see whether the new admissions policy has an effect on the success rate. A two-tailed test is planned. For what values of y,

the number of successful graduate students in the sample of 20, would you accept H_1, if α is set at 0.05?

a) $y < 10.5$

b) $y > 17.5$

c) $9.5 < y < 18.5$

d) $y < 9.5$ or $y > 18.5$

e) $y < 10.5$ or $y > 17.5$

f) $10.5 < y < 17.5$.

6. H_o: $\mu = 100$ is tested against H_1: $\mu \neq 100$ by selecting a sample of n = 100. The two critical values of the sample mean expressed in terms of z are determined as -2.576 and +2.576 for $\alpha \leqq .01$. The value of the test statistic in the sample is $z_{obs} = -2.63$. What conclusion do you draw?

a) Reject H_o, since $z_{obs} < -2.576$.

b) Reject H_o, since $z_{obs} > -2.576$.

c) Fail to reject H_o, since $z_{obs} < -2.576$.

d) Fail to reject H_o, since $z_{obs} > -2.576$.

e) Impossible to say without further information.

7. A standardized ability test with mean = 100 and standard deviation = 10 is administered to a sample of 220 students in a particular school district to determine whether the students in this district also have a mean of 100 or whether their mean score is less than 100. (Assume that the standard deviation does not change.) The sample mean score is found to be 98.3. What is the one-tailed p-value of this result?

a) 0.006 d) 0.068

b) 0.432 e) 0.994.

c) 0.494

8. It is known that the mean systolic blood pressure for men between 18 and 25 years of age is 122, with a standard deviation of 11. A U.S. army medical officer measures the systolic blood pressure of 200 soldiers in this age range to determine whether it is the same as that of the general population (H_o: $\mu = 122$) or different (H_1: $\mu \neq 122$). The mean systolic blood pressure in his sample is 124. What is the *p*-value of this result?

a) 0.005 d) 0.495

b) 0.010 e) 0.857

c) 0.429 f) 0.990.

9. Your local chamber of commerce has asked you to test whether the proportion of single-item sales above $100 made to customers that are 25 years or younger is the same as it was in 1969 (H_o: $p = .2$) or whether it has increased (H_1: $p > .2$). You are instructed to restrict the probability of a Type I error to 1%. In a random sample of 25 single-item sales above $100, you find that eleven of the customers are 25 years or younger. What conclusion do you draw?

a) There is not sufficient evidence to reject H_o: $p = .2$ (p-value < .01).

b) There is not sufficient evidence to reject H_o: $p = .2$ (p-value > .01).

c) There is sufficient evidence for H_1: $p > .2$ (p-value < .01).

d) There is sufficient evidence for H_1: $p > .2$ (p-value > .01).

10. The null hypothesis that the mean IQ score of male college students is the same as that of female college students is tested against the alternative hypothesis that their mean IQ scores differ. (The population standard deviation of the IQ scores of both male and female college students is known to be 10.) The mean IQ score in a random sample of 15 male students is 113; in a random sample of 19 female students the mean IQ score is 115. Compute the absolute value of the test statistic, z_{obs}, for this test.

a) 0.17

b) 0.58

c) 1.68

d) 1.83

e) Cannot be determined without further information.

Analysis

Correct Answer		Explanation of Wrong Answers
1b	a	you computed y_{crit} for $\alpha/2$, as for a two-tailed test;
	c	you exceeded $\alpha = 0.01$;
	e	you computed y_{crit} for H_1: $p > 0.7$; always sketch H_o and H_1;
	d	see answers c and e;
	f	see answers a and e.
2a	b	is the acceptance region for H_o, not H_1;
	c	you used the normal table incorrectly: α is a tail probability, $P(z > z_{crit})$, which you have to subtract from 0.5 to find the probability listed in the table, $P(z < z_{crit})$;
	f	you thought that H_1 is $\mu < 69\%$, but it is $\mu > 69\%$;
	e	see answers b and f;
	d	see answers c and f.
3c	a e	you incorrectly matched the columns of p for the upper cumulative and the rows of y for the lower cumulative in Table A2, or vice versa;
	d	you computed the critical values for $\alpha \leq 0.20$; in a two-tailed test, divide α by 2 for each tail;
	b	see answers a and d.
4e	b	is an intermediate result, i.e., the critical values of z; these have to be converted into $\bar{y}$;

Correct Answer	Explanation of Wrong Answers
	c you have to apply the central limit theorem when converting $\underline{z}_{crit}$ into $\bar{\underline{y}}_{crit}$, i.e., multiply by $\sigma/\sqrt{\underline{n}}$ instead of σ;
	f you computed the critical values for $\alpha \leqq 0.02$; in a two-tailed test, divide a by 2 for each tail;
	a see answers b and f;
	d see answers c and f.
5d	a b are the acceptance regions for $\underline{H}_1$ for one-tailed tests: a) for $\underline{H}_1$: $\underline{p} < 0.7$ b) for $\underline{H}_1$: $\underline{p} > 0.7$; however, a two-tailed test is specified in the question;
	c is the acceptance region for $\underline{H}_o$;
	e is the acceptance region for $\underline{H}_1$ for $\alpha \leqq 0.10$; in a two-tailed test, divide α by 2 for each tail;
	f see answers c and e.
6a	b is the correct conclusion, but for the wrong reason (see Study Guide Tip C);
	c see Objective 22;
	d see Objective 22 and Study Guide Tip C;
	e all relevant information is provided: $\underline{H}_o$, $\underline{H}_1$, $\underline{z}_{crit}$, $\underline{z}_{obs}$.
7a	b you have to apply the central limit theorem-- i.e., divide $(\bar{\underline{y}} - \mu)$ by $\sigma/\sqrt{\underline{n}}$ instead of σ;

Correct Answer	Explanation of Wrong Answers
	c you used the normal table incorrectly: the p-value is a tail probability.
	d see answers b and c;
	e is 1.0 - p-value; you should have made a sketch of the H_o and H_1 distributions and the p-value.
8b	a is the p-value for a one-tailed test; in a two-tailed test, multiply $P(\bar{y} > \bar{y}_{obs})$ by 2 to obtain the p-value of the observed result;
	e see answer 7b;
	f see answer 7c;
	c see answers a and e;
	d see answers a and f.
9c	a b d see Objective 31.
10b	a you did not take the square root of the denominator of the formula for z_{obs};
	d you did not square σ_1 and σ_2 in the formula for z_{obs};
	c see answers a and d;
	e all relevant information is provided: H_o, H_1, $\bar{y}_1$, $\bar{y}_2$, σ_1, σ_2, n_1, and n_2.

CHAPTER 11. *t*-TESTS

OBJECTIVES

Chapter 11 introduces some widely used statistical tests, the *t*-tests. You should be familiar with

I the *t*-distribution;

II the *t*-test of one mean;

III the *t*-test of two independent means;

IV the *t*-test of two dependent means.

You should also be familiar with

V the basics of experimental design.

More specifically, after reading the text you should be able to

I K (1) define the terms *t-distribution*, *degrees of freedom* (Text 11.1);

C (2) distinguish between the *t*-distribution and the normal distribution (Text 11.1);

(3) explain why the *t*-distribution is more variable than the normal distribution (Text 11.1);

(4) distinguish between the formulas for z_{obs} and t_{obs} (Text 11.6);

(5) decide when to use a t-test vs. a z-test (Text 11.1);

A (6) use the t-table (Table A6) to determine

a) the ith percentile of the t-distribution with given degrees of freedom (backward problem) (Text 11.1; Study Guide Tip A);

b) the probability of a given t (forward problem) (Text 11.1; Study Guide Tip A);

II A (7) compute t_{obs} for the test of one mean (Text 11.2);

(8) determine the degrees of freedom and the critical value(s) of t for the t-test of one mean (Text 11.2);

(9) carry out a one- or two-tailed t-test of one mean using the rejection region procedure or the p-value procedure (Text 11.2);

III K (10) list the similarities and differences between the z-test of two independent means and the t-test of two independent means (Text 11.4);

A (11) compute t_{obs} for the test of two independent means (Text 11.4);

(12) determine the degrees of freedom and the critical value(s) of t for the t-test of two independent means (Text 11.4);

(13) carry out a t-test of two independent means for two groups of equal or unequal sample size using the rejection region procedure or the p-value procedure (Text 11.4);

IV K (14) list the similarities and differences between the t-test of two independent means and the t-test of two dependent means (Text 11.5);

A (15) compute t_{obs} for the test of two dependent means (Text 11.5);

(16) determine the degrees of freedom and the critical value(s) of t for the t-test of two dependent means (Text 11.5);

(17) check the reasonableness of a computed value of $\bar{d}$, the mean of the difference scores: $\bar{d} = \bar{y}_1 - \bar{y}_2$ (Text 11.5);

(18) carry out a t-test of two dependent means using the rejection region procedure or the p-value procedure (Text 11.5);

V K (19) define the terms manipulated variable, observed variable, irrelevant variable, randomized variable, matched variable, independent-groups design, dependent-groups design, related-groups design, correlated-groups design, matched-groups design, repeated-measures design (Text 11.3);

(20) state the basic principle of experimental design (Text 11.3);

(21) list three qualifications to the general statement of the superiority of the dependent-groups design (Text 11.3);

(22) list three ways to minimize the effect of an irrelevant variable (Text 11.3);

C (23) distinguish between an independent-groups design and a dependent-groups design (Text 11.3);

(24) distinguish independent- and dependent-groups designs from independent and dependent variables (Text 11.3; Study Guide Tip B);

(25) for a given experiment identify the manipulated variable, the observed variable, and irrelevant variables (Text 11.3);

(26) decide between an independent-groups design and a dependent-groups design for a given experiment (Text 11.3);

A (27) use a random number table to randomly assign

a) subjects to one of two (or more) conditions (Text 11.3);

b) one member of each pair of matched subjects to one of two conditions (Text 11.3).

Chapter 11 t-TESTS

TIPS AND REMINDERS

A. Use of the t-table

The *t*-table (A6) differs from the normal table (A3) in several ways:

normal table	t-table
--percentiles in margin of table;	--percentiles in body of table;
--probabilities in body of table;	--probabilities at top of table;
--same values, regardless of sample size;	--different values for different sample sizes (degrees of freedom in margin of table);
--large number of percentiles and their probabilities are listed;	--only selected percentiles and their probabilities are listed;

Therefore, the use of the *t*-table in solving forward and backward problems differs from that of the normal table.

Backward problems, e.g., determining the critical value of *t* for a given α: Select the column of the *t*-table that is headed by the appropriate probability (i.e., $t_{.050}$ for $\alpha \leqq .05$ in a one-tailed test); select the row with the appropriate degrees of freedom (as determined from the sample size and the type of test). The value in the body of the table at the intersection of row and column is the desired t_{crit}.

Forward problems, e.g. computing the *p*-value for a given t_{obs}: Here a sketch of the problem is very useful. Since the *t*-table only lists selected percentiles of the *t*-distribution, most often you will not find the exact

Figure SG 11.1 Obtaining a p-value

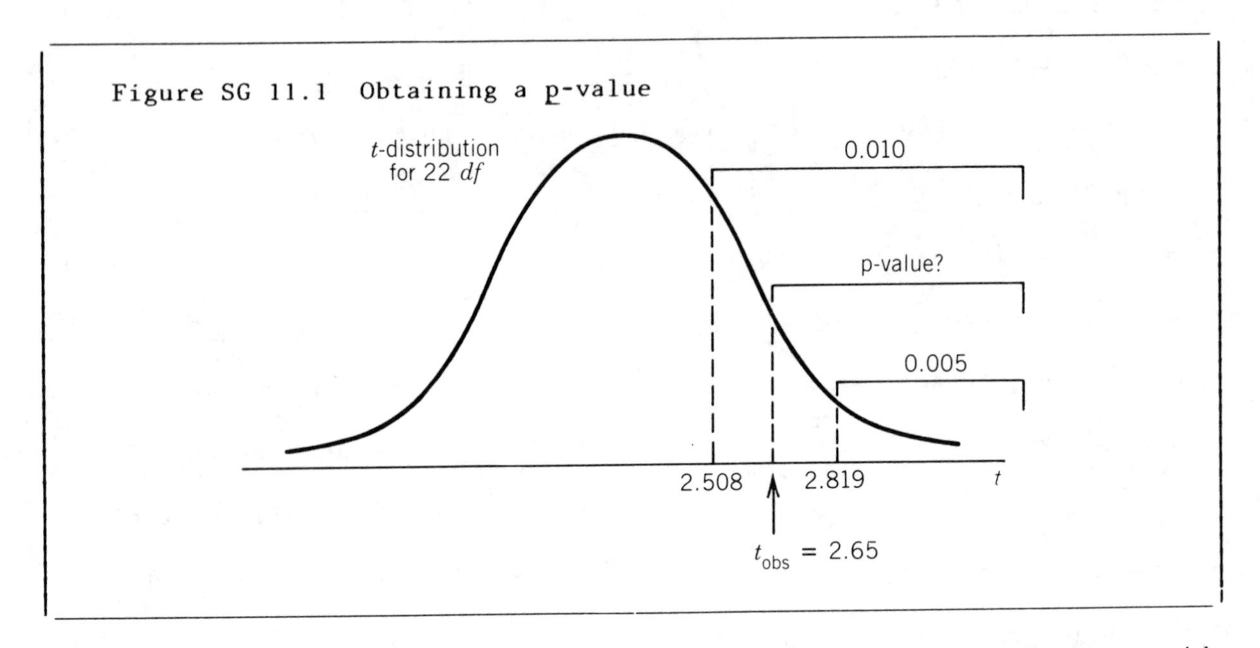

value of the given t_{obs} in the table. However, in the row of the t-table with the appropriate degrees of freedom, you can locate a value of t that is larger than but close to t_{obs}, and a value that is smaller than but close to t_{obs}. Indicate the relative locations of these two values of t and of t_{obs} on a sketch of the t-distribution (see Figure SG 11.1).

The p-value of t_{obs} is the probability of obtaining t_{obs} or a more extreme result, which is represented in the figure by the area in the tail of the t-distribution to the right of $t_{obs} = 2.65$ (assuming a one-tailed test). Note, that this area is larger than the area to the right of $t = 2.819$ which is .005 according to the t-table, but smaller than the area to the right of 2.508 which is .010. Since p-values are compared to conventional levels of α, a p-value is reported as "less than" the conventional level of α that just exceeds it. For this example you would report: *p*-value < .010. If α was specified as .01 or a larger value, the *p*-value would lead to rejecting H_o.

Though it is a true statement that in this example the *p*-value > .005, it is not a useful statement. The inequality, *p*-value > .005, is true for *p*-values as different as .007, .01, .12, or even .65. While some of these *p*-values lead to rejecting H_o for $\alpha \leq .01$, others would not. In contrast, all values for which *p*-value < .010 is true would lead to rejecting H_o.

When the *p*-value of an observed result exceeds the specified value of α, the *p*-value reported as "greater than" α. If, in this example, α was specified as .005, the *p*-value would be reported as *p*-value > .005. In this case, the actual size of the *p*-value is not important since all *p*-values that are greater than .005 would lead to a failure to reject H_o.

A sketch will help you to keep all these different considerations in mind and to use the *t*-table intelligently.

B. Distinguishing "Independent" and "Dependent"

An independent-groups design has

1. Independent Variable
   (either experimentally-defined
   or naturally-occurring)
2. Dependent Variable
   (unrelated or independent
   for different subjects)
3. Irrelevant Variables
   (held constant or randomized)

A dependent-groups design has

1. Independent Variable
   (experimentally-defined)
2. Dependent Variable
   (related or dependent
   for pairs of subjects)
3. Irrelevant Variables
   (one or more are matched;
   others are held constant
   or randomized)

Chapter 11 t-TESTS

PRACTICE QUIZ

Questions

1. For 10 degrees of freedom, what is the probability that t_{obs} will exceed 2.1 if the null hypothesis is true?

   a) < 0.100 d) < 0.010

   b) < 0.050 e) < 0.005

   c) < 0.025 f) > 0.100.

2. A sample of 25 language aptitude test scores has a mean of 75 and a standard deviation (s) of 7. A language school wants to test whether or not this sample comes from a population with a mean larger than 72. Compute t_{obs}, the observed value of the test statistic.

   a) 0.086 d) -0.086

   b) 0.429 e) -0.429

   c) 2.143 f) -2.143.

3. A professor wants to determine whether or not the score of a sample of 20 statistics students comes from a population with $\mu = 83$. If $\alpha \leqq 0.05$, for what values of t should he accept the null hypothesis?

   a) $-1.729 < t < 1.729$

   b) $-1.960 < t < 1.960$

   c) $-2.093 < t < 2.093$

   d) $t < -1.729$ or $t > 1.729$

   e) $t < -1.960$ or $t > 1.960$

   f) $t < -2.093$ or $t > 2.093$.

4. Given $\underline{H}_o$: $\mu = 19$, $\underline{H}_1$: $\mu \neq 19$, and a sample of size 13 with $\bar{y} = 22$ and $\underline{s} = 3.9$, compute the p-value of the observed result:

a) p-value > .01

b) p-value > .02

c) p-value > .10

d) p-value < .01

e) p-value < .02

f) p-value < .10.

5. The biology aptitude test scores of two samples of science students are summarized by $\bar{y}_1 = 90$, $\underline{s}_1 = 6$, $\underline{n}_1 = 20$ and $\bar{y}_2 = 78$, $\underline{s}_2 = 8$, $\underline{n}_2 = 15$. Compute the absolute value of $\underline{t}_{obs}$, the value of the test statistic for the two-tailed test of the null hypothesis that the two samples come from the same population.

a) 13.42

b) 5.13

c) 5.08

d) 4.97

e) 1.73

f) .073.

6. The manufacturing process of a papermill has been modernized. Two samples of 12 quality-control test scores each--one taken for the old process and one for the new process--are to be compared. What values of $\underline{t}$ constitute the rejection region for the null hypothesis that there is no difference in mean quality between the two samples, assuming a two-tailed test and $\alpha \leqq .05$?

a) $-1.796 < t < 1.796$

b) $-2.074 < t < 2.074$

c) $-2.201 < t < 2.201$

d) $t < -1.796$ or $t > 1.796$

e) $t < -2.074$ or $t > 2.074$

f) $t < -2.201$ or $t > 2.201$.

7. Given H_o: $\mu_1 = \mu_2$, H_1: $\mu_1 \neq \mu_2$, two random samples of size $n_1 = 7$ and $n_2 = 12$, with $t_{obs} = 2.63$, what conclusion should be drawn for this t-test of two means, following the p-value procedure?

a) Reject H_o (p-value < .02).

b) Reject H_o (p-value < .01).

c) Reject H_o (p-value < .005).

d) Fail to reject H_o (p-value < .02).

e) Fail to reject H_o (p-value < .01).

f) Fail to reject H_o (p-value < .005).

8. A group of eight sophomores were given an English achievement test before and after receiving instruction. Their scores are presented below.

Student	Before	After
A	20	28
B	18	22
C	17	15
D	16	17
E	14	18
F	14	20
G	12	9
H	9	7

Compute the absolute value of $\underline{t}_{obs}$, the test statistic for the one-tailed test of the hypothesis that instruction increases English achievement test scores.

a) 0.336
b) 0.487
c) 0.739
d) 1.378
e) 2.444
f) 2.677.

9. Consider a $\underline{t}$-test of two dependent means with hypotheses $\underline{H}_o$: $\mu_d = 0$ and $\underline{H}_1$: $\mu_d \neq 0$. Thirteen sets of paired scores have $\underline{t}_{obs} = 2.204$. For a test with $\alpha \leqq .05$ which conclusion should be drawn?

a) Reject H_o, because t_{obs} falls into the rejection region.

b) Reject H_o, because $t_{obs} < t_{crit}$.

c) Fail to reject H_o, because t_{obs} falls into the acceptance region.

d) Fail to reject H_o, because $t_{obs} > t_{crit}$.

e) Impossible to say without knowing the actual scores.

10. Refer to Question 8. Suppose that t_{obs} for another sample of eight students is determined as 2.83. What is the p-value of this result?

a) > .025 d) < .01

b) < .025 e) > .05

c) > .01 f) < .05

Analysis

Correct Answer	Explanation of Wrong Answers
1b	a c d e f see Study Guide Tip A for the importance of a sketch.
2c	a you divided $(\bar{y} - \mu)$ by $s(\sqrt{n})$ instead of by $s/\sqrt{n}$;
	b you should have applied the central limit theorem;
	f the numerator of the t_{obs} formula is $(\bar{y} - \mu)$, not $(\mu - \bar{y})$;
	d see answers a and f;
	e see answers b and f.
3c	a is the acceptance region for $\alpha \leq 0.10$;
	b you should have used the t-table instead of the normal table;
	f is the rejection region;

Correct Answer	Explanation of Wrong Answers
	d see answers a <u>and</u> f;
	e see answers b <u>and</u> f.
4e	b see Study Guide Tip A;
	c you should have applied the central limit theorem;
	d is the one-tailed p-value; however, a two-tailed test is specified by H_1;
	a see answers b <u>and</u> d;
	f see answers b <u>and</u> c.
5c	e you omitted the expression $\sqrt{1/n_1 + 1/n_2}$ in the denominator of the t_{obs} formula;
	a you did not square s_1 and s_2 in the s_p formula;
	d you applied the wrong s_p formula (for equal n);
	f you incorrectly used s_p^2 instead of s_p in the t_{obs} formula;
	b see answers a <u>and</u> f.
6e	f you used the df of the test of two dependent means (n - 1) instead of (n_1 + n_2 - 2);
	d you made two errors: you used $\alpha \leq 0.10$ (two-tailed); also see answer f;
	b is the acceptance region;
	c see answers f <u>and</u> b;
	a see answers d <u>and</u> b;

Correct Answer	Explanation of Wrong Answers
7a	b is the correct conclusion, but based on a one-tailed $\underline{p}$-value; however, a two-tailed test is specified by $\underline{H}_1$;
	c is the correct conclusion, but incorrect $\underline{p}$-value; see Study Guide Tip A;
	d the $\underline{p}$-value is correct, but the wrong conclusion is drawn; review Text 11.4;
	e wrong conclusion (see Text 11.4); one-tailed $\underline{p}$-value though two-tailed test is specified by $\underline{H}_1$;
	f wrong conclusion (see Text 11.4); wrong $\underline{p}$-value; see Study Guide Tip A.
8d	a you incorrectly used $\underline{s}_d^2$ instead of $\underline{s}_d$ in the $\underline{t}_{obs}$ formula;
	b you should have applied the central limit theorem;
	c you used the <u>wrong test</u>: two independent means instead of two dependent means;
	e you probably incorrectly summed the absolute values of $\underline{d}$ ($\Sigma\lvert\underline{d}\rvert$) instead of the values of $\underline{d}$ ($\Sigma\underline{d}$) in the $\underline{s}_d$ formula;
	f you used the number of observations as $\underline{n}$ instead of the number of <u>pairs</u> of scores.
9a	b c d see Objective 18;
	e all relevant information is provided: $\underline{H}_o$, $\underline{H}_1$, $\underline{n}$, α, and $\underline{t}_{obs}$.

Correct Answer	Explanation of Wrong Answers
10b	a c d e see Study Guide Tip A; f is the two-tailed p-value; however, a one-tailed test is specified.

CHAPTER 12. ESTIMATION

OBJECTIVES

Chapter 12 introduces you to the second of the two main techniques for making a statistical inference. To refresh your memory, inferences about a population are made from observations of a random sample drawn from that population (see P6: The Basic Principle of Statistical Inference in Text 6.4). One technique of inference making, hypothesis testing, was discussed in Chapters 9 and 10. Another approach, estimation, is introduced now. You should be familiar with

I point estimates;

II interval estimates;

III the first method for planning sample size.

More specifically, after reading the text, you should be able to

I K (1) define the terms point estimate, unbiased estimate, biased estimate, error of the estimate, standard error of the estimate (Text 12.3);

C (2) distinguish between a biased and an unbiased estimate of a population parameter (Text 12.2);

(3) explain why $\underline{\bar{y}}$ is an unbiased estimate of μ (Recall the central limit theorem) (Text 12.3);

(4) distinguish between the adjusted and the unadjusted sample variance as an estimate of the population variance (Text 12.2);

(5) distinguish between computing a standard error when σ is known and estimating the standard error when σ itself has to be estimated by $\underline{s}$ (Text 12.3);

A (6) estimate the standard error of

(a) $\underline{\bar{y}}$ (Text 12.3);

(b) $\underline{\bar{y}}_1 - \underline{\bar{y}}_2$ (Text 12.4);

(c) $\underline{\bar{d}}$ (Text 12.4);

(d) $\underline{\hat{p}}$ (Text 12.5);

(e) $\underline{\hat{p}}_1 - \underline{\hat{p}}_2$ (Text 12.5);

(7) apply the principle of usefulness to the estimates of the standard error (Text 12.3);

II K (8) define the terms <u>interval estimate</u>, <u>confidence interval</u>, <u>confidence coefficient</u> (Text 12.3);

(9) list the two differences between the formulas for the limits of a confidence interval when σ is known vs. when σ is unknown (Text 12.3);

C (10) interpret the confidence coefficient in terms of the probability that the interval includes the true population parameter (Text 12.3);

(11) explain why the confidence limits for proportions are only approximations (Text 12.5):

A (12) find the limits of a $(1 - \alpha)100\%$ confidence interval of a population parameter by

(a) selecting the appropriate point estimate;

(b) computing or estimating its standard error;

(c) determining the appropriate percentile of the z- or t-distribution;

(Text 12.3, 12.4, 12.5; Study Guide Tip C);

(13) calculate the width of a $(1 - \alpha)100\%$ confidence interval (Text 12.3);

(14) apply the empirical rounding rules of Text Section 12.5 concerning the accuracy of estimated confidence limits for proportions;

III C (15) describe the relationship between the standard error (SE) and sample size (n) given that σ or its estimate s remains constant (Text 12.3, 12.6; Study Guide Tip B);

A (16) determine the sample size needed for a desired standard error of estimate given a known or estimated population standard deviation (Text 12.6).

Chapter 12 ESTIMATION

TIPS AND REMINDERS

A. Point Estimates

To be valid a point estimate should be

(a) unbiased, i.e., its mean error, computed from all possible random samples should be zero, or at least it should be as unbiased as possible (e.g., choose s over s' as a point estimate for σ);

(b) normally distributed, or at least approximately normally distributed (e.g., $\hat{p}$).

To be useful, a point estimate should also be

(c) precise, i.e., the variability of the errors should be small. The standard error of the estimate (SE) serves as a measure of this variability: it is the standard deviation of the errors in the estimate, computed in all possible random samples.

B. Interval Estimates

The width of a confidence interval (CI)--see formulas in Text Table 12.4--is determined by the confidence coefficient, $(1 - \alpha)100\%$, and the standard error (SE).

Increasing the confidence coefficient (e.g., from 95% to 99%) increases the width of the CI because the percentile of the relevant normal or t-distribution is larger for an α-value of .01 than for .05 (e.g. $z_{.005}$ = 2.576 vs. $z_{.025}$ = 1.960). This also has an intuitive interpretation: with equal SE, we have to have a wider interval to be 99% rather than only 95% sure that the interval includes the true value of the parameter.

Decreasing the value of the SE will decrease the width of the interval. As you can see from the formulas for est(SE) in Text Table 12.4, SE decreases as the sample size, n, increases. This also has an intuitive interpretation: we can have greater faith in the precision of a point estimate from a larger sample; thus we do not need as wide a confidence interval.

C. Percentile For Given Confidence Coefficients

Text Table 12.4 is the basic summary of the formulas for point estimates, standard errors, and confidence intervals. Each formula for the confidence interval includes a percentile of the normal or the t-distribution, corresponding to the $(1 - \alpha)100\%$ confidence coefficient. This percentile is, in fact, the same as the critical value of a hypothesis test at that α (two-tailed). This correspondence is explained further in Text C.1. For present purposes, we show below the general formula for each percentile, an example, and a reference to the hypothesis test.

One mean

Percentile: $t^{n-1}_{\alpha/2}$

Example: n = 25, confidence coefficient = 95%

$$t^{24}_{.025} = 2.064$$

Hypothesis test: test of one mean (Text 11.2)

Chapter 12 ESTIMATION

Two independent means

Percentile: $t_{\alpha/2}^{n_1+n_2-2}$

Example: $n_1 = 30$, $n_2 = 32$, confidence coefficient = 99%

$t_{.005}^{60} = 2.660$

Hypothesis test: test of two independent means (Text 11.4)

Two dependent means

Percentile: $t_{\alpha/2}^{n-1}$

Example: n = 25 (pairs), confidence coefficient = 90%

$t_{.05}^{24} = 1.711$

Hypothesis test: test of two dependent means (Text 11.5)

One proportion

Percentile: $z_{\alpha/2}$

Example: n = 43, confidence coefficient = 95%

$z_{.025} = 1.960$

Hypothesis test: binomial test (Text 9.1, 10.1, 9.4)

Two independent proportions

Percentile: $z_{\alpha/2}$

Example: $n_1 = 100$, $n_2 = 80$, confidence coefficient = 99%

$z_{.005} = 2.576$

Hypothesis test: test of two independent proportions (Text 14.3)

D. Forward and Backward Problems

Recall the notion of forward and backward problems discussed in Text Chapter 7 (forward: the computation of a percentile from a given score, and backward: the determination of the score which has given percentile). Analogously, you can look at the computation of a standard error for the estimate (SE) from a given sample as a forward problem, and the first method for planning sample size as its associated backward problem (where you specify the desired SE, know or estimate the population standard deviation, and compute the necessary sample size). Thus, mathematically, the two tasks are the two sides of the same coin. Conceptually, however, the backward problem of planning sample size is a step beyond the techniques for statistical analysis that you have encountered in previous chapters (including the computation of a SE) into the domain of experimental design. You will encounter more of this in Chapter 13. For the time being you may find it useful to think of experimental design as a backward problem to the forward problem of statistical analysis.

E. Symbol Exercise

Match each symbol or equation in the left column with the expression that best describes it in the right column. The solutions are listed at the end of the chapter.

______	1. $\Sigma yP(y)$	a.	lower limit for the $(1 - \alpha)$ 100% confidence interval with σ known
______	2. σ^2	b.	sample proportion, y/n
______	3. $t^{1}_{.10}$	c.	90th percentile of t-distribution with 1 degree of freedom
______	4. $\dfrac{\Sigma(y - \bar{y})^2}{n - 1}$	d.	upper limit for the $(1 - \alpha)$ 100% confidence interval for μ with σ unknown
______	5. $\bar{y} - z_{\alpha/2} \cdot SE$	e.	$\Sigma(y - \mu)^2 P(y)$
______	6. $\dfrac{\Sigma(y - \bar{y})^2}{n}$	f.	$\sqrt{\dfrac{\hat{p}\hat{q}}{n}}$
______	7. $\bar{y} + t^{n-1}_{\alpha/2} \cdot est(SE)$	g.	$s/\sqrt{n}$
______	8. $\hat{p}$	h.	μ
______	9. $SE_{\bar{y}}$	i.	s'^2, biased estimate of σ^2
______	10. $est(SE_{\bar{y}})$	j.	$\sqrt{\dfrac{pq}{n}}$
______	11. $SE_{\hat{p}}$	k.	$\sigma/\sqrt{n}$
______	12. $est(SE_{\hat{p}})$	ℓ.	s^2, unbiased estimate of σ^2.

Chapter 12 ESTIMATION

PRACTICE QUIZ

Questions

1. A developmental psychologist wishes to determine the average amount of time required to complete a problem solving task which has been designed for preschoolers. She presents the problem to a random sample of 8 children in this age group, and obtains the following results (in seconds):

   14, 9, 11, 17, 15, 17, 12, 16.

   What is the estimated standard error of the mean completion time for the population?

a)	0.4	d)	2.9
b)	1.0	e)	3.1
c)	2.7	f)	13.9.

2. A pharmaceutical company wants to know the difference in mean absorption time of two different drugs in muscle tissue. Eighteen specimens of muscle tissue are used: nine are injected with Drug A and nine with Drug B. Absorption (in minutes) is measured as follows:

Drug A	Drug B
2.1	1.5
1.9	1.9
2.3	2.1
1.7	1.6
1.8	1.6
1.7	1.8
2.2	2.0
2.5	1.9
1.7	1.4

What is the standard error of the estimate of the difference in muscle tissue absorption time between the two drugs?

a) 0.02 d) 0.08

b) 0.03 e) 0.13

c) 0.06 f) 0.23.

3. The department of transport does a study to determine what proportion of adults living in urban centers regularly uses public transit systems. In a random sample of 211 adults, 127 report regular use of transit systems. Estimate the standard error of the estimate for the population proportion of transit users that is suggested by this sample.

a) 0.00 d) 0.60

b) 0.03 e) 8.74.

c) 0.04

4. Samples are drawn from two populations, A and B. 90% confidence intervals for the means of both populations are reported as follows:

   $22 < \mu_A < 30$ $\quad\quad$ $43 < \mu_B < 47.$

   From this information, we can conclude that

   a) we can be more confident that the reported interval for μ_A contains the true mean than we can for the reported interval for μ_B.

   b) we can be more confident that the reported interval for μ_B contains the true mean than we can for the reported interval for μ_A.

   c) we can be equally confident for both means.

   d) it is impossible to determine relative confidence from the above data.

5. An experiment was conducted to determine factors which might affect scores on a verbal-reasoning test. It was thought that there might be a difference between students who had taken at least one logic course and those who had not taken any logic course. In order to control for difference in IQ, each of 6 students selected as having at least one logic course was paired with a student of equal IQ who had not had a logic course. Scores on the verbal-reasoning tests were then obtained for all 12 students. The 6 paired observations are as follows:

Pair	Logic	No Logic
1	49	55
2	40	22
3	32	18
4	70	60
5	83	76
6	55	51

Find the 95% confidence limits for difference in mean score on the verbal-reasoning test for students with at least one logic course and students with no logic course.

a) -19.0 to 34.7 d) 1.1 to 14.6

b) - 1.0 to 16.6 e) 4.2 to 11.4

c) - 0.6 to 16.2 f) 38.2 to 55.8.

6. In a public opinion poll about a particular issue, 490 of 700 male and 64 of 100 female respondents indicated a favorable attitude. Estimate the difference in the population proportions favoring the issue by using a 99% confidence interval.

a) -.06 to .18

b) -.07 to .19

c) -.10 to .23

d) -.1 to .2.

7. An experiment was conducted to compare the severity of side effects (measured on a scale of 1 to 10) of two antipsychotic drugs. Fifty schizophrenic patients are randomly divided between the two drugs (two equal groups); side effects are measured after the drugs have been taken for 2 weeks. The means and variances of the side-effects measure for the two samples were

$\bar{y}_1 = 7.5,\ s_1^2 = 0.08$ and $\bar{y}_2 = 8.3,\ s_2^2 = 0.20,$

respectively. What is the width of the 99% confidence interval for the difference in side effects?

a) 0.11 d) 0.30

b) 0.15 e) 0.57

c) 0.21 f) 0.80.

8. The confidence interval for the mean age of death for women in North America was determined as CI_f: 63.5 - 73.5 from a sample of 600 women. In a sample of 500 men, the confidence interval for the same parameter was CI_m: 59.5 - 69.5. For which confidence interval could you be more confident that it contained the true population parameter (i.e., which has a larger confidence coefficient)?

a) CI_m, because it is based on a smaller $\underline{n}$.

b) CI_f, because it is based on a larger $\underline{n}$.

c) CI_m, because the mean age of death for men is lower.

d) CI_f, because the mean age of death for women is higher.

e) Equally confident for both, because the width of the two intervals is the same.

f) Impossible to say without further information.

9. An IQ test has been standardized to have a standard deviation of 15 in its target population of high school students. The Ministry of Education wants an estimate of the mean IQ of Ontario high school students with a standard error not exceeding 0.5. What sample size is needed?

a) 8 c) 56

b) 30 d) 900.

10. You are asked to give an estimate of the difference in mean lifetime of rats on two different diets. To improve accuracy, the rats will be matched on their level of fitness. A pilot study indicated that the standard deviation of the differences is 12 days. The standard error of your estimate is not to exceed 1.2 days. How many pairs of rats would you use?

a) 5 d) 100

b) 10 e) 200

c) 50 f) 207.

Analysis

Correct Answer		Explanation of Wrong Answers
1b	f	this is the point estimate of the mean completion time in the population, not its standard error;
	a	you forgot to take the square root of n;
	d	this is the adjusted sample standard deviation; to obtain the standard error, divide it by $\sqrt{n}$;
	e	you used the sample variance instead of the standard deviation in your computation of $SE_{\bar{y}}$;
	c	you computed the unadjusted sample variance; see Study Guide Tip A(a); also, see e.
2e	b	you used the pooled sample variance instead of the s.d. in your computation of $SE_{\bar{y}_1-\bar{y}_2}$;
	d	you computed the standard error for the estimate of the difference between two dependent groups; this is, however, an independent groups design;
	a	see d; also, you made a mistake when using the formula for $SE_{\bar{y}_1-\bar{y}_2}$: you used s_d^2 instead of s_d;
	c	you correctly computed the pooled sample s.d., but then divided it by the square root of the number of observations, as if estimating $SE_{\bar{y}}$ instead of $SE_{\bar{y}_1-\bar{y}_2}$;
	f	you reported $\bar{y}_1 - \bar{y}_2$, not the standard error.

Correct Answer	Explanation of Wrong Answers
3b	a you forgot to take the square root of $\hat{p}\hat{q}/n$;
	c e you treated this question like an estimate of a population mean: $SE_{\hat{p}} = \sqrt{\frac{\hat{p}\hat{q}}{n}}$, not c $\hat{p}/\sqrt{n}$, or e $y/\sqrt{n}$;
	d this is the point estimate of the population proportion, not its standard error.
4c	a b d the confidence coefficient, which is 90% for both intervals, indicates the probability that the interval will contain the true population parameter.
5b	a you computed the 95% confidence limits for the difference between two independent groups; this is, however, a dependent groups design;
	c you used the wrong t-distribution: n instead of $n - 1$ degrees of freedom;
	d you should have used $t^{n-1}_{\alpha/2}$ instead of $z_{\alpha/2}$;
	e you forgot to take the square root of n;
	f $\bar{d}$, not Σd, is the point estimate for the difference in mean score.

Correct Answer	Explanation of Wrong Answers
6b	a a confidence interval corresponds to a two-tailed test (as explained in Text Chapter C): therefore you have to use $z_{\alpha/2}$, not z_{α}, in your computation of the confidence limits;
	c to obtain $SE_{\hat{p}_1-\hat{p}_2}$ you have to take the square root of the sum of $\hat{p}_1\hat{q}_1/n_1$ and $\hat{p}_2\hat{q}_2/n_2$, not the sum of the square roots;
	d since the smallest number of successes or failures in either group is 36, i.e., greater than 20, the limits are accurate to two decimal places.
7e	c you used the pooled sample variance instead of the s.d.;
	f this is the point estimate of the difference, not the width of the 99% confidence interval;
	d this is half the width of the desired interval; you may have obtained this result by not multiplying $(t_{\alpha/2}^{n_1+n_2-2} \cdot s_p\sqrt{2/n})$ by two, or, by treating this question like an estimate of one mean, i.e., $2 \times (t_{\alpha/2}^{n-1} \cdot s_p/\sqrt{n})$, where n = total number of patients;
	a see d and c;
	b you treated this question like an estimate of one mean (see d), and also did not multiply $(t_{\alpha/2}^{n-1} \cdot s_p/\sqrt{n})$ by two to obtain the width of the interval.

Correct Answer	Explanation of Wrong Answers
8f	a b c d e see Objective 15: since the two samples come from different populations, it is necessary to know the size of the two standard deviations before any other statement about the relative sizes of the confidence coefficients can be made.
9d	a b c you solved the equation for est($SE_{\bar{y}}$) incorrectly for _n_; see Study Guide Chapter 1, Tip (10): $n = (s/SE_{\bar{y}})^2$, _not_ a $s \cdot SE_{\bar{y}}$, b $s/SE_{\bar{y}}$, c $(s \cdot SE_{\bar{y}})^2$.
10d	b f you solved the equation for est($SE_{\bar{d}}$) incorrectly for _n_; see Study Guide Chapter 1, Tip (10): $n = (s_d/SE_{\bar{d}})^2$, _not_ b $s_d/SE_{\bar{d}}$, f $s_d \cdot SE_{\bar{d}}$; e you incorrectly treated the question like an _independent_ instead of a dependent groups design; c the _n_ you obtain _is_ the number of pairs to be used; you should not divide this number by two; a see b and c.

Symbol Exercise Solutions

1h, 2e, 3c, 4ℓ, 5a, 6i, 7d, 8b, 9k, 10g, 11j, 12f.

CHAPTER 13. POWER AND EFFECT SIZE

OBJECTIVES

Chapter 13 is a step into the domain of experimental design. Should you ever be involved in some research of your own, this chapter will be one of the most important ones to remember. Estimating the power of an experiment, before executing it and analyzing the data, is very informative (it tells you, for example, whether the sample size you are planning to use will give you sufficient power; also see Text C.4). Unfortunately, many researchers have not yet acquired the habit of estimating power. You should be familiar with

I the notion of the power of a statistical test;

II methods for hypothesizing an unknown effect size;

III the second method for planning sample size.

More specifically, after reading the text, you should be able to

I K (1) define the terms <u>power</u>, <u>hypothesized</u> <u>effect</u> <u>size</u> (<u>HES</u>), δ (Text 13.1, 13.2);

(2) list the five factors which affect the power of a test (Text 13.1);

(3) specify the HES for (Text 13.1)

(a) a test of one mean;

(b) a test of one proportion;

(c) a test of two independent means;

C (4) explain why a specific alternative hypothesis is needed to compute the power of a test (Text 13.1);

(5) explain, by means of a diagram, the effect on the power of a test of increasing/decreasing (Text 13.1);

(a) HES,

(b) σ,

(c) $\underline{n}$,

(d) α;

(e) switching from a one-tailed to a two-tailed test;

(6) summarize the effects of HES, σ, and $\underline{n}$, as the effect of δ on power (Text 13.2);

A (7) compute the power of a test

(a) from first principles (Text 9.3, 9.4, 10.6, 13.1);

(b) from the approximate power table A7 (Text 13.2);

(8) apply your knowledge of the relationship between HES, σ, $\underline{n}$, α, one- or two-tailed test, and power when planning an experiment: e.g., planning sample size (see Objective 13, Text 13.4), or, deciding between a dependent- or independent-groups design (Text 13.1);

II K (9) define the term <u>Cohen's</u> <u>Scale</u> (Text 13.3);

C (10) interpret the ratio of <u>HES</u> to σ as an indication of the size of the effect (visually, as the amount of overlap between the population distributions under the two hypotheses)(Text 13.3);

A (11) estimate the expected effect size and the population standard deviation for an experiment

(a) directly, based on previous studies and/or standardized test information (Text 13.3);

(b) as a ratio of <u>HES</u> and σ, also usually based on previous research (Text 13.3);

(c) using Cohen's scale of suggested ratios (Text 13.3);

III C (12) contrast the first and second method for planning sample size (Text 13.4; Study Guide Tip A);

A (13) determine the sample size needed to obtain a desired power in a test of specified α (one- or two-tailed) given the <u>HES</u> and known or estimated σ or an estimate of their ratios (Text 13.4).

TIPS AND REMINDERS

A. Two Methods For Planning Sample Size

The two methods may be summarized by an analogy: estimation is to hypothesis testing as the first method for planning sample size is to the second method.

	First method	Second method
Requirements:	desired SE;	desired power;
	estimated σ.	value of α;
		one- or two-tailed test;
		estimate of HES and σ (or their ratio).
Result:	necessary n.	necessary n.

B. The Importance of a Sketch

The location of the regions representing α, β, and power in the distributions of the test statistic when H_o is true and when H_1 is true depends among other things on whether the H_1 distribution is to the right or to the left of the H_o distribution. To avoid common sources of error (e.g., locating the critical value in the wrong tail of the H_o distribution, or computing the value of β when, in fact, you want to compute the power), always make a rough sketch of the two distributions as in Figure SG 13.1, indicating the regions representing α, β, and power. Such a sketch will also help you check on the reasonableness of any calculated value.

Figure SG 13.1 Sketches for α, β, and Power

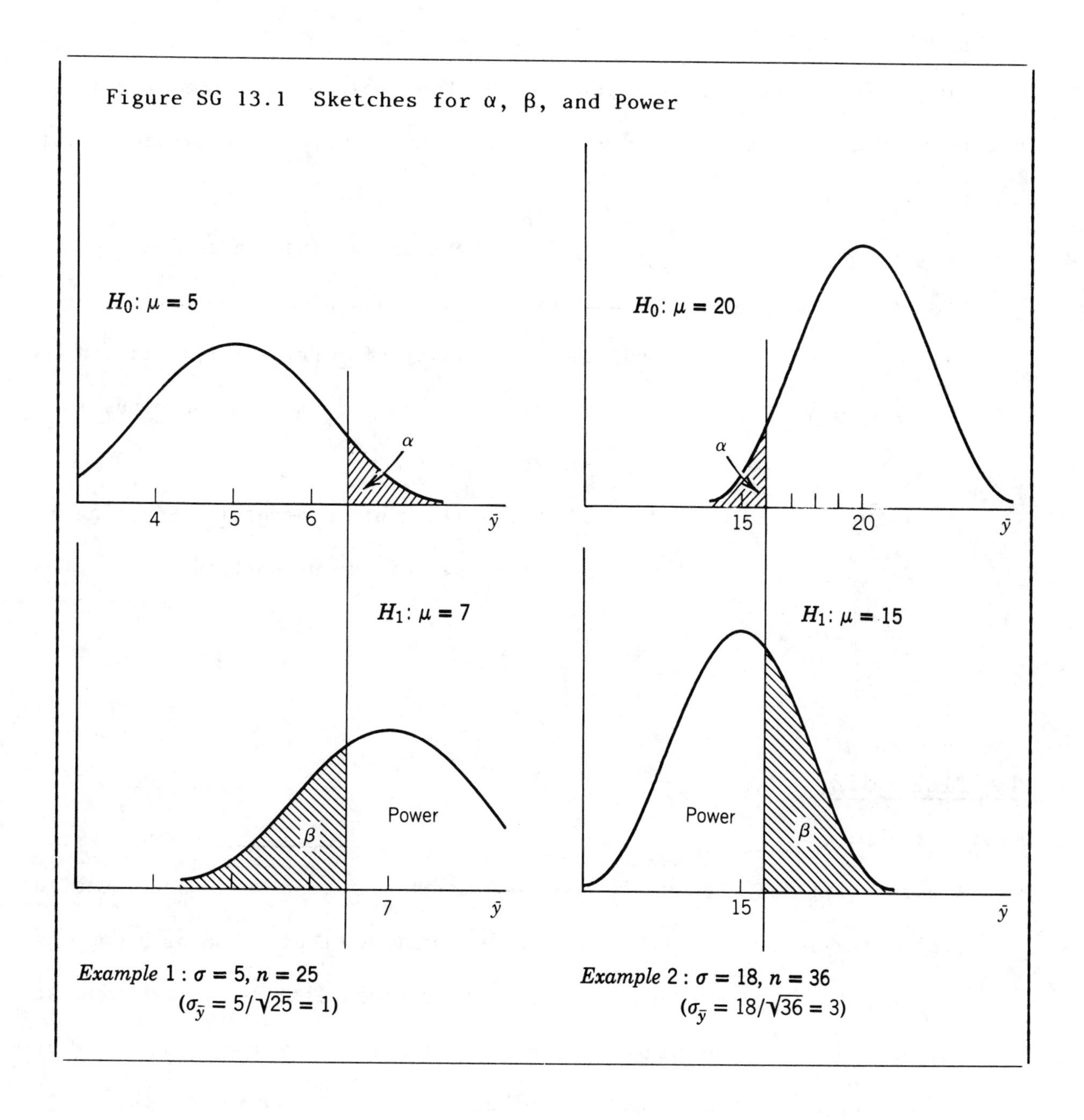

C. Symbol Exercise

Match each symbol or equation in the left column with the expression that best describes it in the right column. The solutions are listed at the end of the chapter.

_____	1. $1 - \beta$	a.	SE for a test of two independent means
_____	2. α	b.	δ for a test of one proportion
_____	3. δ	c.	the probability of correctly rejecting H_o
_____	4. $\lvert\mu_1 - \mu_o\rvert$	d.	the absolute value of mu sub one minus mu sub zero
_____	5. $\sigma\sqrt{2/n}$	e.	the probability of incorrectly rejecting H_o
_____	6. $\sqrt{p_o q_o / n}$	f.	SE for a test of one proportion
_____	7. $\frac{HES}{\sqrt{p_o q_o}} \sqrt{n}$	g.	$\frac{HES}{SE}$

PRACTICE QUIZ

Questions

1. A one-tailed test of two independent means ($\alpha = .05$) is to be conducted with a sample of $n = 300$ people to determine whether a HES of 3 does in fact exist. The new research assistant decides that a two-tailed test would be more appropriate, and she also increases the sample size. What effect do these two changes together have on the power of the test?

a) power increases

b) power decreases

c) power stays the same

d) impossible to say without further information.

2. An investigator estimates the <u>HES</u> for a test of one mean to be of a certain size and estimates the population standard deviation to be 8. A two-tailed test (α = .01) is to be conducted on a sample of 45 rats. After reviewing some previous studies, he realizes that the <u>HES</u> is twice as large as he originally thought, and that σ is in fact 16. How does this change the power of the proposed test?

   a) power increases

   b) power decreases

   c) power stays the same

   d) impossible to say without further information.

3. A chewing gum manufacturer claims that its new product will hold its flavor significantly longer than other brands. Measurements on a large number of other brands show that the flavor lasts for 7 minutes on the average. Test the alternative hypothesis that the flavor of the new brand will last for 8.5 minutes. Assume that the standard deviation is 2 in either case. For a sample size of 25 specimens of the new brand, compute the critical value of the test statistic (the mean length of time the flavor lasts in a sample of 25). The probability of a Type I error is specified as .01.

a) $\bar{y}_{crit} = 2.35$

b) $\bar{y}_{crit} = 6.07$

c) $\bar{y}_{crit} = 7.93$

d) $\bar{y}_{crit} = 11.65$.

4. It is claimed that the mean distance that students travel to university is 14 km. We wish to test the hypothesis that the mean distance is 14 km (H_o) against the hypothesis that the mean distance is 14.5 km (H_1). Assume that the standard deviation is 2 km in either case.
Let the test statistic be the mean distance travelled by a random sample of 25 students. The probability of a Type I error is set to 0.01 so that $\bar{y}_{crit} = 14.93$. Compute the power for this test.

   a) .01

   b) .14

   c) .86

   d) .99.

5. A certain poison seems to kill about 20% of the laboratory animals that receive it. You are asked to test whether the proportion of deaths is the same for a new mutant of albino mice (H_o: $p = .20$) or whether it is lower, i.e., H_1: $p = .15$. α is specified as .05 (one-tailed), and you are provided with a random sample of 50 mice. What is the probability that you will reject H_o, if the true proportion of deaths in the mice population is in fact .15?.

a) .06 d) .64

b) .15 e) .71

c) .23 f) .88.

6. A labor economist plans a sample survey of male workers in a given occupation to see if the mean weekly wage differs from the national average of $200. He plans a test of one mean using a sample size of 100 with a non-directional significance criterion of $\alpha = 0.01$. He expects that the occupation has a mean wage differing about $1 from the national average and that the standard deviation of wages in that occupation is $4. What is the power of the test he is planning?

a) .01 d) .57

b) .02 e) 2.50.

c) .47

7. The difference in life expectancy between married and single males is to be examined. The two hypotheses are $\underline{H}_o: \mu_m - \mu_s = 0$ and $\underline{H}_1: \mu_m - \mu_s = 5$ years. Assume that the standard deviation is 11 years in both populations. Compute the value of δ for random samples of 50 married and 50 single men if the probability of a Type I error is to be 0.01 (one-tailed).

a) .45 d) 2.27

b) .48 e) 4.55

c) .99 f) 5.00.

8. Rats raised under standard conditions gain 80 grams in weight in the first 100 days. An experiment is planned where the rats will be raised in special conditions to see whether their mean weight gain will differ from 80 grams, in either direction. It is desired to detect the difference if it is greater than one-quarter of a standard deviation with a probability of .90, using a 0.05 level test. What is the necessary sample size?

   a) 4
   b) 12
   c) 13
   d) 135
   e) 169.

9. What sample size, per group, would be needed in the following experiment: two independent groups with α = 0.01 one-tailed, with a desired probability of getting a significant result = 0.95 if the true difference between the groups is one-quarter of a standard deviation.

   a) 14
   b) 29
   c) 256
   d) 282
   e) 512
   f) 564.

10. How many days, selected at random, would be needed to determine whether the proportion of rainy days in Vancouver is equal to the Canadian national average (H_o: p = .10) or whether it is higher or lower by .05? Use α = .05 and a desired power of .95.

   a) 467
   b) 380
   c) 32
   d) 26.

Chapter 13 POWER AND EFFECT SIZE

Analysis

Correct Answer	Explanation of Wrong Answers
1d	a b c even though the effects of the two changes are in opposite directions (one increases, the other decreases the power), you need to know the new sample size before a statement about the overall effect can be made.
2c	a b d all relevant information is provided: the formula for δ for a test of one mean is $(\underline{HES}/\sigma)\sqrt{\underline{n}}$. If both HES and σ are doubled, the factor of 2 can be cancelled in this formula and δ, and thus power, remain the same.
3c	b you located the critical value in the wrong tail of the distribution of $\bar{y}$ (see Text 9.3 and Study Guide Tip B);
	d you failed to apply the central limit theorem in your computation for the critical value (see Text 9.3);
	a see b and d.
4b	a you used μ_o instead of μ_1 in the computation of the power (see Text 9.3; Study Guide Tip B); you in fact computed α;
	c this is β, which is 1 − power;
	d see a; also, this is $1 - \alpha$.
5c	b this would be the correct power if the test were two-tailed rather than one-tailed; make sure you select the correct column of Table A7;

Correct Answer		Explanation of Wrong Answers
	f	this is δ; the power is obtained from Table A7 using δ and the specified value of α;
	a	this is SE, not the power (see Objective 7b);
	d	$SE = \sqrt{p_o q_o / n}$, not $\sqrt{p_o p_1 / n}$;
	e	you forgot to take the square root of $p_o q_o$ in your computation of δ.
6c	d	this is the power for a one-tailed test;
	e	this is δ; the power is obtained from Table A7, using δ and the specified value of α;
	a	$\delta = (HES/\sigma)\sqrt{n}$, not (HES/σ);
	b	see a and d.
7d	b	this is the power of the test; you were asked to compute δ;
	e	you computed δ as if for a test of one mean; this is a test of two independent means;
	c	see a and e;
	a	$\delta = (HES/\sigma)\sqrt{n/2}$, not simply HES/σ;
	f	this is HES, not δ.
8e	d	you used the value of δ that would correspond to the specified power and a one-tailed test; make sure to always select the correct column of Table A7;

Correct Answer	Explanation of Wrong Answers
	c you used the specified value of the power as δ, i.e., you skipped Table A7; you also may have obtained this result by incorrectly solving $\delta = (HES/\sigma)\sqrt{n}$ for _n_: $n = (\delta\sigma/HES)^2$, _not_ $(\delta\sigma/HES)$;
	a you probably made _both_ mistakes described in c;
	b see d and second part of c.
9e	f you used the value of δ that would correspond to the specified power and α in a _two_-tailed test;
	c the _n_ that you solve for _is_ the sample size per group; you should not divide the obtained value by two;
	d see c and f;
	b you used the specified value of the power as δ, i.e., you skipped Table A7;
	a see b and c.
10a	b you used the value of δ that would correspond to the specified power and α in a one-tailed test;
	c you used the specified value of the power as δ, i.e., you skipped Table A7;
	d you incorrectly solved $\delta = (HES/\sqrt{p_o q_o})\sqrt{n}$ for _n_; $n = \delta^2 p_o q_o/(HES)^2$, not $\delta^2 p_o/HES$.

Chapter 13 POWER AND EFFECT SIZE

Symbol Exercise Solutions

1c, 2e, 3g, 4d, 5a, 6f, 7b.

REVIEW CHAPTER C

OBJECTIVES

Four important aspects of hypothesis testing and estimation are discussed in this chapter. The discussion serves to review Chapters 9 to 13. You should be familiar with

I the relationship between hypothesis testing and estimation;

II the classification of the procedures studied so far;

III the requirements for the validity of statistical procedures;

IV the relationship between statistical significance and effect size.

More specifically, after reading the text you should be able to

I C (1) explain the relationship between estimation and hypothesis testing (Text C.1);

A (2) carry out a test of a hypothesis at level α (two-tailed) when you are given the $(1 - \alpha)100\%$ confidence interval (Text C.1);

(3) make an interval estimate when you are given a point estimate, the sample sizes, and the value of the test statistic (Text C.1);

II A (4) select the appropriate procedure for a test or estimate (Text C.2);

III K (5) list and describe the four requirements for the validity of statistical procedures (Text C.3);

C (6) decide whether a given requirement applies to a particular procedure (Text C.3);

IV C (7) explain why the finding and reporting of a statistically significant result is not sufficient without further information (Text C.4);

(8) explain the conditions under which the null hypothesis can be accepted (Text C.4);

A (9) compute the effect size from the value of a test statistic (Text C.4);

(10) determine whether the null hypothesis can be accepted in a given test (Text C.4).

TIPS AND REMINDERS

A. Symbol Exercises--Review of Chapter 9 to 13

Here are four symbol exercises. For each one, match each symbol or expression in the left column with the expression that best describes it in the right column. The solutions are listed at the end of the chapter.

Chapter C REVIEW

Exercise I

_____	1. $\bar{y}_{crit}$	a.	adjusted standard deviation
_____	2. H_o	b.	the test statistic in a t-test
_____	3. p-value	c.	$\sqrt{pq/n}$
_____	4. s	d.	biased estimate of σ^2
_____	5. t_{obs}	e.	the null hypothesis
_____	6. s_p	f.	the probability of the observed statistic or more extreme result
_____	7. s'^2	g.	pooled standard deviation
_____	8. est$(SE_{\bar{y}})$	h.	the critical value for the sample mean
_____	9. $SE_{\hat{p}}$	i.	$s/\sqrt{n}$

Exercise II

_____	1. α	a.	delta
_____	2. β	b.	sigma sub y-bar
_____	3. δ	c.	alpha
_____	4. μ	d.	sigma
_____	5. σ	e.	beta
_____	6. $\sigma_{\bar{y}}$	f.	mu

Exercise III

_____	1. α	a.	population mean
_____	2. β	b.	probability of a Type II error

_____	3. δ	c.	population standard deviation
_____	4. μ	d.	probability of a Type 1 error
_____	5. σ	e.	standard error of the sample mean
_____	6. $\sigma_{\bar{y}}$	f.	HES/SE

Exercise IV

_____	1. z_{obs}	a.	$\sigma/\sqrt{n}$
_____	2. H_1	b.	number of degrees of freedom
_____	3. s'	c.	the test statistic in a *z*-test
_____	4. df	d.	hypothesized effect size
_____	5. s_d	e.	unadjusted standard deviation
_____	6. s^2	f.	$\sqrt{\frac{\hat{p}\hat{q}}{n}}$
_____	7. $SE_{\bar{y}}$	g.	the alternative hypothesis
_____	8. est($SE_{\hat{p}}$)	h.	standard deviation of differences
_____	9. HES	i.	unbiased estimate of σ^2

Symbol Exercise Solutions

Exercise I: 1h, 2e, 3f, 4a, 5b, 6g, 7d, 8i, 9c.

Exercise II: 1c, 2e, 3a, 4f, 5d, 6b.

Exercise III: 1d, 2b, 3f, 4a, 5c, 6e.

Exercise IV: 1c, 2g, 3e, 4b, 5h, 6i, 7a, 8f, 9d.

CHAPTER 14. TESTS FOR DICHOTOMOUS VARIABLES

OBJECTIVES

Chapter 14 introduces you to a series of tests for dichotomous variables, including median tests which reduce to tests for a dichotomous variable. You should be familiar with

I dichotomous variables;

the following tests for dichotomous variables:

II the sign test;

III the chi-square test of two independent proportions;

IV the test of more than two independent proportions;

the following median tests:

V the test of one median;

VI the test of two or more independent medians.

More specifically, after reading the text you should be able to

I K (1) define the term dichotomous variable (Text 14.1);

C (2) give examples of (a) naturally dichotomous variables, (b) dichotomous variables that arise from a comparison, and (c) dichotomous variables that are reduced numerical variables (Text 14.1);

(3) interpret a test of a dichotomous variable as a test of proportions (e.g., the binomial test) (Text 14.1);

II C (4) decide when the sign test is applicable (Text 14.2);

(5) interpret the sign test as a binomial test (of one proportion) (Text 14.2);

A (6) state $\underline{H}_o$ and $\underline{H}_1$ for the sign test (Text 14.2);

(7) carry out the sign test (Text 14.2):

a) compute the test statistic $\underline{y}$;

b) determine the rejection region for a given α (or compute the $\underline{p}$-value of the test statistic), using either the individual or cumulative binomial table;

III K (8) define the terms <u>observed frequency</u>, <u>expected frequency</u>, <u>chi-square distribution</u> (Text 14.3);

C (9) decide when the test of two independent proportions is applicable (Text 14.3);

(10) interpret the test statistic χ^2_{obs} as an index of the differences between the observed frequencies and their corresponding expected frequencies (Text 14.3; Study Guide Tip D);

(11) explain why the *df* of the appropriate χ^2-distribution is determined from the dimensions of the frequency table (rather than, as previously, from the sample size) (Text 14.3);

(12) explain why only upper-tail percentiles of the χ^2-distribution are required for the rejection region in the χ^2-test (Text 14.3; Study Guide Tip D);

(13) state the relationship between the *z*-test and the χ^2-test of two independent proportions (Text 14.4);

(14) judge whether the approximation of the distributions of the test statistics z_{obs} and χ^2_{obs} to the standard normal and χ^2-distribution, respectively, is satisfactory for a given set of data (Text 14.4);

(15) distinguish between "independence of two variables" and the notions of an "independent variable" and an "independent-groups design" (Text 14.4);

(16) interpret the test of independence of two dichotomous variables as a test of two independent proportions (Text 14.4);

A (17) state H_o and H_1 for the test of two independent proportions (Text 14.3);

(18) carry out the test of two independent proportions in one of two ways (Text 14.3, 14.4):

a) i) summarize the data in a 2 × 2 table of observed frequencies;

ii) compute the expected frequencies $\underline{E}_{ij}$;

iii) compute the test statistic χ^2_{obs};

iv) determine the rejection region for a given α (or compute the $\underline{p}$-value of the test statistic), using the χ^2-table ($\underline{df}$ = 1);

b) i) compute the test statistic $\underline{z}_{obs}$;

ii) determine the rejection region for a given α (or compute the $\underline{p}$-value of the test statistic), using the standard normal table;

(19) state $\underline{H}_o$ and $\underline{H}_1$ for the test of independence of two dichotomous variables (Text 14.4);

(20) carry out the test of independence of two dichotomous variables following the procedure of the chi-square test of two independent proportions (Objective 18) (Text 14.4);

IV C (21) interpret the test of more than two independent proportions as a generalization of the test of two independent proportions (Text 14.5);

A (22) state $\underline{H}_o$ and $\underline{H}_1$ for the test of more than two independent proportions (Text 14.5);

(23) carry out the test of $\underline{J}$ independent proportions (Text 14.5):

a) summarize the data in a 2 × $\underline{J}$ table of observed frequencies;

b) compute the expected frequencies;

c) compute the test statistic χ^2_{obs};

d) determine the rejection region for a given α (or compute the $\underline{p}$-value of the test statistic), using the χ^2-table ($\underline{df}$ = $\underline{J}$ - 1);

V C (24) decide when to use the median test as opposed to other tests of central tendency (Text 14.6);

A (25) state $\underline{H}_o$ and $\underline{H}_1$ for the test of one median (Text 14.6);

(26) carry out the test of one median (Text 14.6):

a) reduce the numerical scores to the dichotomous variable "scores above $\underline{vs}$. below $\underline{M}_o$":

b) perform a sign test on this dichotomous variable (Objective 7);

VI A (27) state $\underline{H}_o$ and $\underline{H}_1$ for the test of the equality of medians of $\underline{J}$ independent populations (Text 14.6);

(28) carry out the test of $\underline{J}$ independent medians (Text 14.6):

a) estimate the overall population median $\underline{M}$ by the overall sample median $\hat{\underline{M}}$;

b) reduce the numerical scores in each group to the dichotomous variable "scores above $\underline{vs}$. below $\hat{\underline{M}}$";

c) summarize the data in a $2 \times \underline{J}$ observed frequency table, whose first and second rows contain the frequency of scores below and above $\underline{\hat{M}}$, respectively, for each group ($\underline{J}$ columns for $\underline{J}$ groups);

d) perform a test of $\underline{J}$ independent proportions (Objective 23) on the $2 \times \underline{J}$ frequency table.

TIPS AND REMINDERS

A. Summary of Statistical Tests

The following pages present a summary of when and how to apply the statistical tests introduced in Chapter 14.

Sign Test

Applicability. The sign test may be applied to data from dependent-groups designs.

Hypotheses. H_o: $P(x_1 > x_2) = 0.5$

H_1: $P(x_1 > x_2) \neq 0.5$,

where x_1 and x_2 are the scores of the first and second member, respectively, of a given pair.

Test statistic. y = number of pairs where $x_1 > x_2$.

Distribution of test statistic under H_o. Binomial with $p = 0.5$ and n = number of subject pairs.

Source of critical values. Tables A1, A2.

Treatment of ties. Drop tied pairs from analysis; decrease n accordingly.

Chi-square Test of Two Independent Proportions

Applicability. This chi-square test may be used when the subjects in two independent groups are scored on a dichotomous variable ("success" and "failure"). The test may also be used when subjects in one group are scored on two dichotomous variables; the test is then a test of the independence of two dichotomous variables.

Hypotheses. H_o: $p_1 = p_2$ (or, the two variables are independent)

H_1: $p_1 \neq p_2$ (or, the two variables are dependent),

where p_1 and p_2 are the probabilities of "success" in the 1st and 2nd population, respectively.

Test statistic.

$$\chi^2_{obs} = \sum_{ij}\sum \frac{(O_{ij} - E_{ij})^2}{E_{ij}} = \sum_{ij}\sum \frac{O^2_{ij}}{E_{ij}} - n,$$

for 2 × 2 frequency table, where O_{ij} = observed frequency, E_{ij} = expected frequency, i = row indicator, j = column indicator, and n = total sample size. (See Text 14.4 for an alternative test statistic, z_{obs}.)

Distribution of test statistic under H_o. χ^2_{obs} is approximately $\chi^2(1)$; z_{obs} is approximately standard normal. Both approximations are satisfactory if all $E_{ij} \geq 5$.

Source of critical values. Table A8 (for χ^2-distribution); Tables A3, A4 (for standard normal).

Treatment of ties. Not applicable.

Test of More Than Two Independent Proportions

Applicability. This chi-square test may be used when the subjects in $\underline{J}$ groups are scored on a dichotomous variable ("success" and "failure").

Hypotheses. $\underline{H}_o$: $\underline{p}_1 = \underline{p}_2 = \ldots = \underline{p}_J$

$\underline{H}_1$: $\underline{H}_o$ is false,

where $\underline{p}_i$ is the probability of "success" in the $\underline{i}$th population.

Test statistic. χ^2_{obs} for 2 × $\underline{J}$ frequency table. (See chi-square test of two independent proportions, above, for the formulas for χ^2_{obs}.)

Distribution of test statistic under $\underline{H}_o$. χ^2_{obs} is approximately $\chi^2(\underline{J} - 1)$.

Source of critical values. Table A8.

Treatment of ties. Not applicable.

Test of One Median

Applicability. This test is used to test the central tendency of numerical scores in one group. Some of the scores, at the extremes of the distribution of scores, may be numerically indeterminate, but these scores can still be compared with the hypothesized median.

Hypotheses. H_o: $M = M_o$

H_1: $M \neq M_o$

where M_o is a given, known, number (the hypothesized median). These hypotheses reduce to:

H_o: $p = 0.5$

H_1: $p \neq 0.5$,

where p is the probability that a score is greater than M_o.

Test statistic. y = number of scores which are greater than M_o.

Distribution of test statistic under H_o. Binomial with $p = 0.5$ and n = number of subjects.

Source of critical values. Tables A1, A2.

Treatment of ties. Drop scores which are exactly equal to M_o; decrease n accordingly.

Test of Two or More Independent Medians

Applicability. This test is used to test the equality of the central tendency of numerical scores in J groups. Some of the scores, at the extremes of the distribution of scores, may be numerically indeterminate, but the overall sample median can still be calculated.

Hypotheses. H_o: $M_1 = M_2 = \ldots = M_J$

H_1: H_o is false,

where M_i is median of the ith population. These hypotheses reduce to:

H_o: $p_1 = p_2 = \ldots = p_J$

H_1: H_o is false,

where p_i is the probability that a score in the ith population is greater than the overall population median.

Test statistic. χ^2_{obs} for $2 \times J$ frequency table. The frequencies are the number of scores, in each group, which are above and below the overall sample median. (See chi-square test of two independent proportions, above, for the formulas for χ^2_{obs}.)

Distribution of test statistic under H_o. χ^2_{obs} is approximately $\chi^2(J - 1)$.

Source of critical values. Table A8.

Treatment of ties. Drop scores which are exactly equal to the overall sample median; decrease n accordingly.

B. Double Summation Notation

The formula for the computation of the test statistic χ^2_{obs} contains two summation symbols ("sigma"):

$$\chi^2_{obs} = \sum_i\sum_j \frac{(O_{ij} - E_{ij})^2}{E_{ij}}.$$

This means that you have to sum the expression $(O_{ij} - E_{ij})^2/E_{ij}$ over all columns j for each row i. To illustrate, let us call the expression $(O_{ij} - E_{ij})^2/E_{ij}$ simply a_{ij}. Then,

$$\chi^2_{obs} = \sum_i\sum_j a_{ij}.$$

Suppose that there are 2 rows (r = 2) and 3 columns (c = 3). To obtain χ^2_{obs}, you have to sum a_{ij} over all 3 columns for each of the 2 rows:

$$\chi^2_{obs} = \sum_{i=1}^{2} \sum_{j=1}^{3} a_{ij}$$

$$= \sum_{i=1}^{2} (a_{i1} + a_{i2} + a_{i3})$$

$$= (a_{11} + a_{12} + a_{13}) + (a_{21} + a_{22} + a_{23}).$$

C. Equivalence of the Two χ^2_{obs} Formulas

It can be shown that the two formulas for χ^2_{obs} given in Text 14.3 and 14.4 are equivalent. The formulas are:

$$\chi^2_{obs} = \sum_i\sum_j \frac{(O_{ij} - E_{ij})^2}{E_{ij}} \quad \text{(definitional formula)}$$

$$\chi^2_{obs} = \sum_i\sum_j \frac{O^2_{ij}}{E_{ij}} - n \quad \text{(computational formula).}$$

<u>We know that in general</u>: $(a - b)^2 = a^2 - 2ab + b^2$

$\Sigma\Sigma(a + b) = \Sigma\Sigma a + \Sigma\Sigma b$

$\Sigma\Sigma(ka) = k \cdot \Sigma\Sigma a$, where <u>k</u> is a constant.

<u>Then</u>:

$$\sum_{ij}\sum \frac{(O_{ij} - E_{ij})^2}{E_{ij}} = \sum_{ij}\sum \frac{(O_{ij}^2 - 2O_{ij}E_{ij} + E_{ij}^2)}{E_{ij}}$$

$$= \sum_{ij}\sum \left\{\frac{O_{ij}^2}{E_{ij}} - 2O_{ij} + E_{ij}\right\}$$

$$= \sum_{ij}\sum \frac{O_{ij}^2}{E_{ij}} - 2\sum_{ij}\sum O_{ij} + \sum_{ij}\sum E_{ij}$$

$$= \sum_{ij}\sum \frac{O_{ij}^2}{E_{ij}} - 2n + n \quad (\text{since } \sum_{ij}\sum O_{ij} = n \text{ and } \sum_{ij}\sum E_{ij} = n)$$

$$= \sum_{ij}\sum \frac{O_{ij}^2}{E_{ij}} - n.$$

D. Rejection Region in Chi-square Tests

Two- sided chi- square tests (e.g., the test of two independent proportions: H_o: $p_1 = p_2$, H_1: $p_1 \neq p_2$) have one-tailed rejection regions, i.e., only in the upper tail of the appropriate χ^2-distribution: $\chi^2_{obs} > \chi^2_{\alpha}(df)$. This is a result of the nature of χ^2_{obs}. χ^2_{obs} is a measure of how much the observed frequencies deviate from the frequencies expected under H_o. When H_o is true, the observed frequencies tend to be close to the expected frequencies; hence, χ^2_{obs} is small. When H_1 is true, the observed and expected frequencies are quite different, but all these differences, whether positive or negative, contribute positively to χ^2_{obs} since each deviation $(O_{ij} - E_{ij})$ is squared in the computation of χ^2_{obs}. Thus, χ^2_{obs} is insensitive to the direction of the observed effect and is always large when H_1 is true.

E. Obtaining a p-value for Chi-square Tests

Since a two-sided H_1 in the chi-square test has a one-tailed rejection region, the p-value of an obtained result, χ^2_{obs}, will simply be the probability of obtaining χ^2_{obs} or a greater value. [Remember that in all other previous two-sided tests (e.g., the t-test), you had to multiply the probability of the obtained or a more extreme result, in one tail, by two in order to get the p-value.]

Figure SG 14.1 is a refresher of the discussion on p-value in Study Guide Chapter 11. A sketch like the one in the figure will help you to determine the p-value for a given χ^2_{obs}. Suppose that $\chi^2_{obs} = 6.51$. Table A8 shows that $\chi^2_{obs} = 6.51$ lies between $\chi^2_{.05}(2) = 5.991$ and $\chi^2_{.025}(2) = 7.378$.

Remember that a p-value is compared to conventional levels of α and is reported as "less than" the conventional level of α that just exceeds it, unless the p-value is greater than the conventional upper limit of α, for example, 0.05 or 0.10. In the latter case, the p-value is reported as "greater than" this conventional upper limit.

For the example in Figure SG 14.1, the p-value does not exceed any such upper limit; it lies between $\alpha = 0.025$ and $\alpha = 0.05$. Since the p-value is reported as "less than" the α level that just exceeds it, we report that p-value < 0.05. Note that χ^2_{obs} is greater than $\chi^2_{.05}(2) = 5.991$, but its p-value is less than 0.05.

Figure SG 14.1 Obtaining a p-value for Chi-square Tests

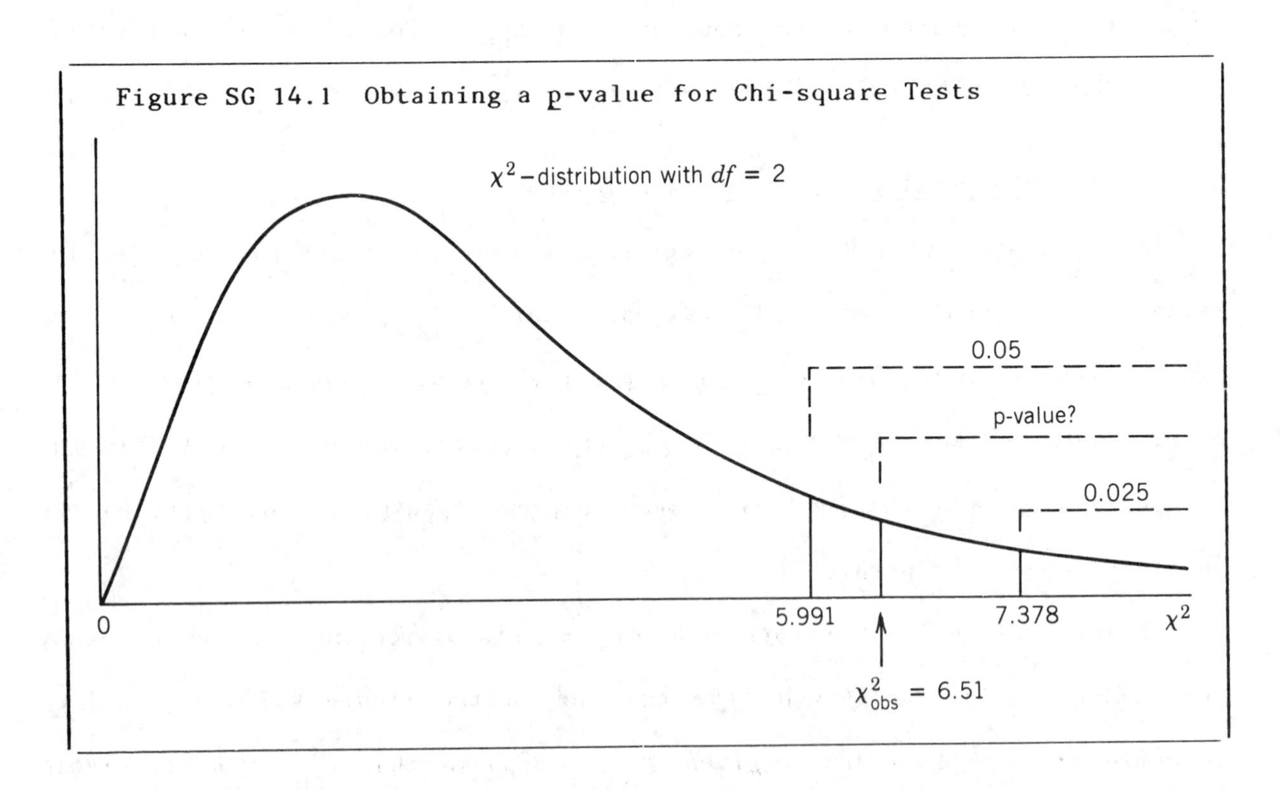

Remember that even though it is true that p-value > 0.025 in this example, this fact is irrelevant for the purpose of statistical inference. Both a p-value of 0.03 and a p-value of 0.40 are "greater than 0.025", yet such p-values would lead to quite different conclusions, since one is less than the conventional upper limit of 0.05 and the other is greater than this limit.

F. Formula Translation

Here is how to say the two formulas for χ^2_{obs}. The definitional formula is

$$\sum_i \sum_j \frac{(O_{ij} - E_{ij})^2}{E_{ij}}.$$

Translated into words, the formula is "the sum, over all rows and columns, of the squared difference between the observed and expected frequencies divided by the expected frequency". (See Tip B.) The computational formula is

$$\sum_i \sum_j \frac{O_{ij}^2}{E_{ij}} - n.$$

Translated into words, the formula is "the sum, over all rows and columns, of the observed frequency squared divided by the expected frequency, minus n, the total number of observations".

G. Symbol Exercise

Match each symbol or equation in the left column with the expression that best describes it in the right column. The solutions are listed at the end of the chapter.

____ 1.	χ^2_{obs}	a.	"Chi-square"
____ 2.	$\hat{M}$	b.	$(z_{\alpha/2})^2$
____ 3.	$\hat{p}$	c.	observed frequency of first column in second row
____ 4.	χ^2	d.	"oh sub i dot", the sum of the ith row
____ 5.	O_{21}	e.	$\sum_i \sum_j \frac{(O_{ij} - E_{ij})^2}{E_{ij}}$
____ 6.	$O_{i.}$	f.	"p hat", the pooled proportion of success
____ 7.	$\chi^2_\alpha(1)$	g.	E_{ij}
____ 8.	$\frac{O_{i.} \times O_{.j}}{n}$	h.	"M hat", the overall sample median

PRACTICE QUIZ

Questions

1. The following table summarizes the observed frequencies in a test of three independent proportions.

Method of Instruction

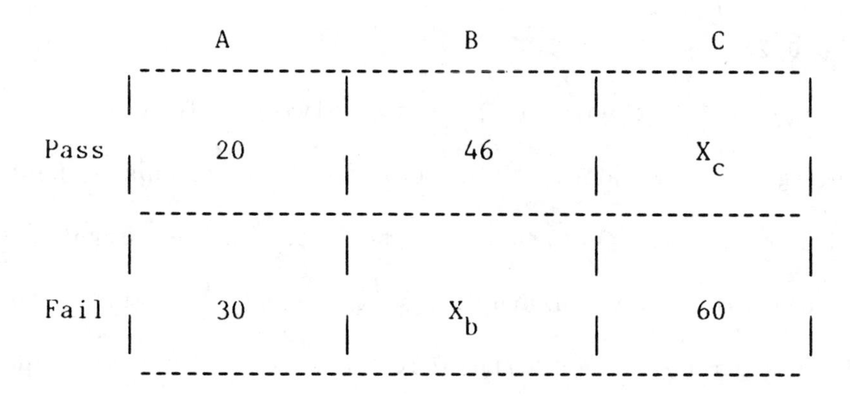

	A	B	C
Pass	20	46	X_c
Fail	30	X_b	60

What are the values of the two unspecified frequencies X_b and X_c?

a) $X_b = 45$; $X_c = 72$

b) $X_b = 56$; $X_c = 50$

c) $X_b = 69$; $X_c = 40$

d) impossible to say without further information.

2. A school board wants to test several programs designed to improve reading skills. Four classes of first-graders are randomly assigned to the four programs (one class to each program). At the end of the year, 19 of the 27 students in program A can read a test-passage error-free, versus 25 of the 32 students in program B, 14 of the 25

students in program C, and 21 of the 29 students in program D. Compute the χ^2_{obs}-statistic for the test of whether the probability of error-free reading differs for the four programs.

a) 121.052
b) 116.415
c) 8.052
d) 3.415
e) 1.052.

3. Professor K. has a new theory of problem solving. To test some of his ideas he assigns 60 students randomly to four groups. Each group receives different instructions on how to solve cryptarithmetic problems. Following the instructions everybody receives the same problem which is to be solved within 10 minutes. The time required to solve the problem is recorded; some people are unable to finish the problem within the 10-minute period.

Which of the following statistical tests should be used to decide whether there is a difference in problem-solving speed among the four groups?

a) sign test
b) test of two independent proportions
c) test of more than two independent proportions
d) test of one median
e) test of two or more independent medians.

4. Eleven middle-aged, male, identical twin pairs take part in a study of the effects of diet on hair loss. One member of each pair is assigned at random to a low-fat diet; the other member receives a normal North American diet. At the end of a two-year period, hair loss is monitored for one week:

	Number of hairs lost	
Pair	Low fat diet	Normal Diet
1	256	242
2	96	70
3	144	170
4	45	52
5	87	59
6	112	98
7	201	185
8	36	21
9	134	100
10	152	137
11	77	64

Compute the *p*-value of the result of this study.

a) .027 d) .064

b) .032 e) .936

c) .054 f) .968.

5. Jill cannot decide whether to settle in city A, B, or C. A friend reminds her to consider age differences in the cities' populations. He just read that the χ^2_{obs}-statistic for the test of differences in median age in these three cities is 6.89, and advises Jill that she should use a .05 level of significance to decide whether the median age differs. What is Jill's conclusion?

a) There *is* evidence for a difference in median age, since $\chi^2_{obs} >$ 3.841.

b) There *is* evidence for a difference in median age, since $\chi^2_{obs} >$ 5.991.

c) There *is* evidence for a difference in median age, since $\chi^2_{obs} <$ 7.378.

d) There is *insufficient* evidence for a difference in median age, since $\chi^2_{obs} > 3.841$.

e) There is *insufficient* evidence for a difference in median age, sinces $\chi^2_{obs} > 5.991$.

f) There is insufficient evidence for a difference in median age, since $\chi^2_{obs} < 7.378$.

6. Two independent samples of subjects ($n_1 = 22$, $n_2 = 25$) are assigned to two different viewing conditions, respectively, for a visual discrimination task. Proportion of success is 0.55 in the first sample, 0.68 in the second sample. What are the z-statistic and rejection region ($\alpha = .05$) for the test of the hypothesis that the proportion of success under the two viewing conditions is the same?

a) $z_{obs} = -0.92$; reject H_o if $z_{obs} < -1.645$ or $z_{obs} > 1.645$.

b) $z_{obs} = -0.92$; reject H_o if $z_{obs} < -1.960$ or $z_{obs} > 1.960$.

c) $z_{obs} = -0.65$; reject H_o if $z_{obs} < -1.645$ or $z_{obs} > 1.645$.

d) $z_{obs} = -0.65$; reject H_o if $z_{obs} < -1.960$ or $z_{obs} > 1.960$.

e) $z_{obs} = -6.51$; reject H_o if $z_{obs} < -1.645$ or $z_{obs} > 1.645$.

f) $z_{obs} = -6.51$; reject H_o if $z_{obs} < -1.960$ or $z_{obs} > 1.960$.

7. A social psychologist claims that taking care of a pet can prolong an old person's life. She distributes pets to fifteen 70-year-old people in a retirement community where the median life expectancy is 74 years. At the end of 7 years, five of her subjects are still alive. The other ten died at the ages of 70, 71, 72, 75, 75, 75, 76, 76, 77, 77, respectively.

   Compute the *p*-value (two-tailed) for this result.

   a) .018 d) .344

   b) .036 e) .500

   c) .172 f) 1.0.

8. A child psychologist believes that first-born children have personality characteristics that are quite different from later-born children. For example, first-born children excel in academic and professional achievement. In order to support his view, the psychologist carries out a study in which the number of first-born and later-born children earning various grades is compared. The study involves using five grade levels (A, B, C, D, E) and determining the number of first-born and later-born children who receive these grades. Therefore, the question is whether the proportion of first-born children differs among the different grade levels.

Which one of the following statistical tests should be applied?

a) sign test

b) test of two independent proportions

c) test of more than two independent proportions

d) test of one median

e) test of two or more independent medians.

9. For a χ^2-test of two independent proportions (H_o: $p_1 = p_2$, H_1: $p_1 \neq p_2$), the test statistic is $\chi^2_{obs} = 4.37$. What conclusion do you draw, if α is specified as .05?

a) Reject H_o, since $\chi^2_{obs} > 0.001$.

b) Reject H_o, since $\chi^2_{obs} > 0.004$.

c) Reject H_o, since $\chi^2_{obs} > 3.841$.

d) Fail to reject H_o, since $\chi^2_{obs} < 5.024$.

e) Fail to reject H_o, since $\chi^2_{obs} < 5.991$.

f) Fail to reject H_o, since $\chi^2_{obs} < 7.378$.

g) Impossible to say without knowing the sample sizes.

10. You are interested to see whether there is a relationship between sex and marital status (single vs. married) for Americans in their thirties. From a large random sample, you compute $z_{obs} = 2.07$. You decide to use $z_{crit} = 1.960$ as the critical value. What can you conclude?

a) Sex and marital status appear to be related.

b) Sex and marital status appear to be unrelated.

c) The proportion of singles seems to be different from men and women.

d) The proportion of singles seems to be the same for men and women.

e) Both a) and c).

f) Both b) and d).

Analysis

Correct Answer	Explanation of Wrong Answers
1d	a b c you need the row or column totals to determine the missing frequencies.
2d	e you forgot to square $(\underline{O}_{ij} - \underline{E}_{ij})$;
	b you forgot to subtract $\underline{n}$ from $\sum_{ij}\sum (O_{ij}^2/\underline{E}_{ij})$ when applying the computational formula;
	c you forgot to square $\underline{O}_{ij}$ in the computational formula;
	a see b and c.
3e	a b c d see Objectives 4, 9, 24.
4d	b you computed the $\underline{p}$-value for a one-tailed test;
	c you computed the probability of $\underline{y}$ being exactly the observed value, instead of being the observed or a more extreme value;
	a see c and b;
	e this is 1 - $\underline{p}$-value;
	f see b and e.

Correct Answer	Explanation of Wrong Answers

5b

a f given your choice of critical value, your conclusion is correct; however, in

- a you used $df = 1$ instead of $df = J - 1$ in finding the critical value;
- f you used $\alpha/2$ instead of α in finding the critical value (see Study Guide Tip D);

c d e you reached the wrong conclusion given your choice of critical value (see Text Chapter 10); also, in

- e your critical value is correct;
- c see f;
- d see a.

6b

d f you used the z_{obs} formula incorrectly:

- d you should have taken the square root of the sum $(\hat{p}_1\hat{q}_1/n_1 + \hat{p}_2\hat{q}_2/n_2)$ instead of taking the square root of each term separately and then taking the sum;
- f you forgot to take the square root of the expression in the denominator;

a c e you forgot to split α for the two-tailed rejection region; also, in

- a z_{obs} is correct;
- c see d;
- e see f.

Correct Answer	Explanation of Wrong Answers
7b	a this is the p-value for a one-tailed test;
	d you have to include the subjects that are still alive in your values for y and n;
	f you did not include the subjects that are still alive in your value for y;
	c see d and a;
	e see f and a.
8c	a b d e see Objectives 4, 9, 24.
9c	b you used the lower tail instead of the upper tail of the χ^2-distribution to find the critical value;
	a see b; you also incorrectly used $\alpha/2$ instead of α (see Study Guide Tip D);
	d you incorrectly used $\alpha/2$ instead of α in finding the critical value (see Study Guide Tip D);
	e you incorrectly computed the df of the critical value as $(r - 1) + (c - 1)$ instead of $(r - 1) \times (c - 1)$;
	f see d and e;
	g the sample sizes are irrelevant for the determination of the df for the χ^2-distribution.

Correct Answer	Explanation of Wrong Answers
10e	a c if the proportion of singles is different for men and women, then sex and marital status are related;
	b d f you reached the wrong conclusion given the observed value and the critical value.

Symbol Exercise Solutions

1e, 2h, 3f, 4a, 5c, 6d, 7b, 8g.

CHAPTER 15. TESTS FOR CATEGORICAL AND RANK VARIABLES

OBJECTIVES

Chapter 15 discusses several tests that will allow you to deal with categorical and rank variables. You should be familiar with

I different types of variables;

II requirements for the validity of χ^2-tests;

the following tests for categorical variables:

III the χ^2-test in an $\underline{r} \times \underline{c}$ table;

IV the test of goodness-of-fit;

V the rank-sum tests for two independent and for two dependent groups.

More specifically, after reading the text you should be able to

I K (1) define the terms <u>numerical variable</u>, <u>categorical variable</u>, <u>rank variable</u> (Text 15.1);

(2) distinguish between (and give examples for)

(a) discrete vs. continuous numerical variables;

(b) naturally vs. reduced categorical variables;

(c) dichotomous vs. polytomous categorical variables;

(d) naturally ranked variables vs. numerical variables reduced to a rank variable;

(Text 15.1; Study Guide Tip A);

II K (3) state the three requirements for the validity of χ^2-tests (Text 15.4; Study Guide Tip C);

III K (4) define the terms <u>homogeneity of J independent distributions</u>, <u>contingency table</u> (Text 15.2);

C (5) interpret the test of equality of two or more independent proportions (Text 14.3, 14.5) as a test of homogeneity of J independent distributions (Text 15.2);

(6) interpret the test of independence of two categorical variables as a special case of the test of homogeneity of J independent distributions (Text 15.2);

A (7) state H_0 and H_1 for the test of homogeneity of J independent distributions (Text 15.2);

(8) carry out the test of homogeneity of J independent distributions (Text 15.2):

(a) summarize the data in an appropriate $\underline{r} \times \underline{c}$ table;

(b) compute the expected frequencies;

(c) compute the test statistic χ^2_{obs};

(d) determine the rejection region for a given α (or compute the $\underline{p}$-value of a given result) from the appropriate χ^2-table [$\underline{df} = (\underline{r} - 1) \times (\underline{c} - 1)$];

IV K (9) define the terms goodness-of-fit, goodness-of-fit test (Text 15.3);

C (10) interpret the binomial test as a goodness-of-fit test for a dichotomous variable (i.e., as a χ^2-test with $\underline{df} = \underline{r} - 1$) (Text 15.3; Study Guide Tip C);

(11) decide when to apply a goodness-of-fit test vs. a test of homogeneity of $\underline{J}$ independent distributions (Text 15.3);

A (12) state $\underline{H}_o$ and $\underline{H}_1$ for the test of goodness-of-fit (Text 15.3);

(13) carry out the test of goodness-of-fit (Text 15.3):

(a) compute the expected frequencies;

(b) compute the test statistic χ^2_{obs};

(c) determine the rejection region for a given α (or compute the $\underline{p}$-value of a given result) from the appropriate χ^2-table ($\underline{df} = r - 1$);

V C (14) contrast the requirements for the $\underline{t}$-test and the rank-sum test as tests of central tendency of two populations (Text 15.5);

(15) decide when to apply a rank-sum test for two independent groups vs. two dependent groups (Text 15.5);

A (16) state H_o and H_1 for the rank-sum test for two independent groups (Text 15.5);

(17) carry out the rank-sum test for two independent groups (Text 15.5):

(a) compute the test statistic W_S;

(b) determine the rejection region for a given α (or compute the p-value of a given result);

(18) state H_o and H_1 for the rank-sum test for two dependent groups (Text 15.6);

(19) carry out the rank-sum test for two dependent groups (Text 15.6):

(a) compute the test statistic T_+;

(b) determine the rejection region for a given α (or compute the p-value of a given result).

TIPS AND REMINDERS

A. Types of Variables

Figure SG 15.1 is a reproduction of Text Figure D.1 and shows the interrelationship between the different types of variables. Table SG 15.1 summarizes this information. As an exercise, think of an example for each of the listed possibilities (a to j) and write it in the example column. You will find one possible set of examples at the end of the chapter.

Figure SG 15.1 Types of Variables

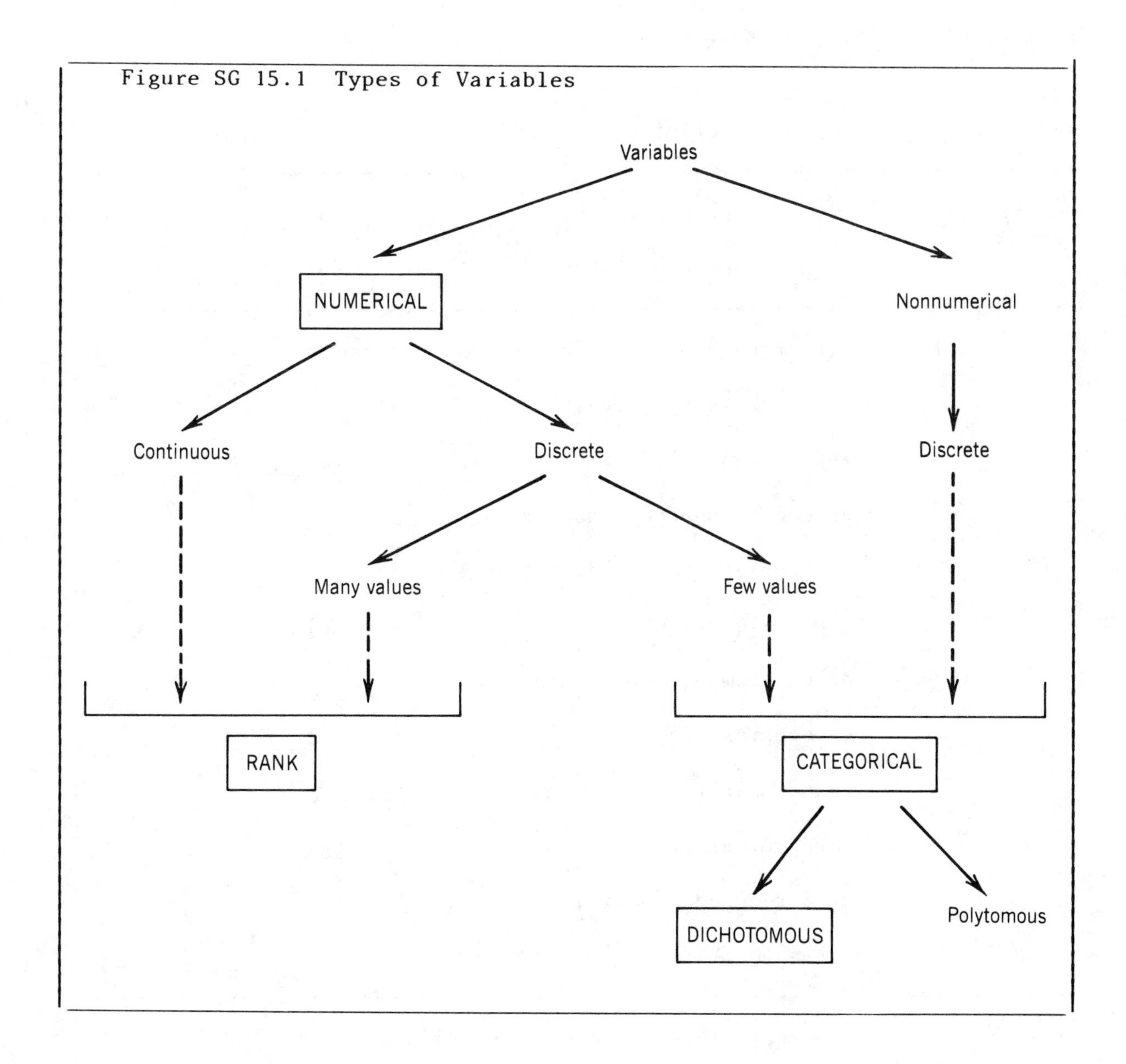

Table SG 15.1 Types of Variables

Type of Variable	Characteristics	Example
numerical	• either discrete (a) or continuous (b)	(a) ______ (b) ______
categorical	• either nonnumerical (c)	(c) ______
	• or discrete numerical with few values (d);	(d) ______
	• depending on the number of categories, called dichotomous (two categories) (e)	(e) ______
	• or polytomous (more than two categories) (f);	(f) ______
	• can be naturally categorical (g)	(g) ______
	• or a reduced numerical variable (h);	(h) ______
rank	a numerical variable reduced to ranks--	
	• either continuous (i)	(i) ______
	• or discrete with many values (j).	(j) ______

B. Summary of Statistical Tests

The following pages present a summary of when and how to apply the statistical tests introduced in Chapter 15.

Chi-square Test in an $r \times c$ Table

Applicability. The chi-square test may be used when subjects in J independent groups are scored on a categorical variable. The test may be used when subjects in one group are scored on two categorical variables; the test is then a test of the independence of two categorical variables.

Hypotheses. H_0: the J population distributions are equal (or, the two variables are independent).

H_1: the J population distributions are not all equal (or, the two variables are dependent).

Test statistic. $$\chi^2_{obs} = \sum_i \sum_j \frac{(O_{ij} - E_{ij})^2}{E_{ij}} = \sum_i \sum_j \frac{O^2_{ij}}{E_{ij}} - n,$$

for an $r \times c$ table, where O_{ij} = observed frequency, E_{ij} = expected frequency, i = row indicator, j = column indicator, and n = total sample size.

Distribution of test statistic under H_0. χ^2_{obs} is approximately $\chi^2[(r - 1) \times (c - 1)]$. The approximation is satisfactory if all $E_{ij} \geq 5$.

Source of critical values. Table A8.

Treatment of ties. Not applicable.

Test of Goodness-of-fit

Applicability. This test is used when testing one categorical variable (r categories) in one population.

Hypotheses. Dichotomous variable: H_o: $p = p_o$, H_1: $p \neq p_o$.

Polytomous variable: H_o: $p_a = p_{ao}$, $p_b = p_{bo}$, . . ., up to r categories. H_1: H_o is false.

Test statistic.

$$\chi^2_{obs} = \sum_i \frac{(O_i - E_i)^2}{E_i} = \sum_i \frac{O_i^2}{E_i} - n,$$

for r categories, where O_i = observed frequency, E_i = expected frequency, i = category indicator, and n = sample size.

Distribution of test statistic under H_o. χ^2_{obs} is approximately $\chi^2(r - 1)$. The approximation is satisfactory if all $E_i \geqq 5$.

Source of critical values. Table A8.

Treatment of ties. Not applicable.

Rank-sum Test For Two Indepdendent Groups

Applicability. This test is used when two independent groups are scored on a numerical variable.

Hypotheses. H_o: the two population distributions are the same.

H_1: the two population distributions are different.

Test statistic. Combine the scores into a single group and assign appropriate ranks. W_S = sum of ranks of the smaller group.

Distribution of test statistic under H_o. The distribution of W_S for the particular (n_S, n_L) combination.

Source of critical values. Table A9.

Treatment of ties. Assign to each of the tied scores the average of the ranks which the scores would have if they were not tied.

Rank-sum Test for Two Dependent Groups

Applicability. This test is used when two matched, or dependent, groups are scored on a numerical variable.

Hypotheses. H_o: the two population distributions are the same.

H_1: the two population distributions are different.

Test statistic. Compute n differences, $d_i = y_{1i} - y_{2i}$; rank absolute differences $|d_i|$, starting with smallest absolute difference (ranked as 1); affix a plus or minus sign to each rank so that the sign of the rank agrees with the sign of the original difference, d_i; compute T_+ = sum of the postive signed ranks.

Distribution of test statistic under H_o. The distribution of T_+ for the given n.

Source of critical values. Table A10.

Treatment of ties. Zero difference (i.e., scores are the same for a matched pair: $d_i = 0$): drop each such pair and reduce n accordingly. Tied difference (i.e., two or more absolute differences, $|d_i|$, are tied): assign to each such difference the average of the ranks which the absolute differences would have if they were not tied.

C. How to Choose the Correct Test

As you are now learning how to do a larger number of statistical tests, the question of "which test when" becomes more and more important. Chapter D of the Text will help you to integrate all previous tests into a common framework. In the meantime, Table SG 15.2 categorizes all the tests that you have encountered in Chapters 14 and 15, except the sign test and the goodness-of-fit test.

How do you pick a test? Ask yourself two questions:

(1) What types of variables are the dependent and the independent variable? In Chapters 14 and 15, the independent variable is either a dichotomous or a categorical (polytomous) variable. The dependent variable is either a dichotomous, categorical, numerical, or rank variable. The particular combination of dependent and independent variable narrows your choice of test to one of the cells in Table SG 15.2.

(2) What requirements are met?

(a) random sampling is absolutely necessary for any statistical inference test;

(b) independence of observations is a necessary requirement for χ^2-tests;

(c) large expected frequencies ($E_{ij} \geqq 5$ in each cell) are necessary for χ^2-tests; if violated, collapse cells (i.e., categories of the independent or dependent variable) to increase the expected cell frequencies.

Table SG 15.2 Categorization of Tests in Chapters 14 and 15

Type of Dependent Variable	Type of Independent Variable	
	Dichotomous	Categorical
Dichotomous	1. two independent proportions (14.3, 14.4)	1. more than two independent proportions (14.5)
Categorical		1. tests in an $\underline{r} \times \underline{c}$ table (15.2)
Numerical	1. two independent medians (14.6)	1. more than two independent medians (14.6)
Rank	1. rank-sum test for two independent groups (15.5) 2. rank-sum test for two dependent groups (15.6)	

There are also practical considerations for test selection. For example, to perform a binomial test for values of n or p which are not in readily available tables, you can resort to the normal approximation to the binomial (Text 9.4) or to the χ^2-test of goodness-of-fit (Text 15.3).

D. Symbol Exercise

Match each symbol or equation in the left column with the expression that best describes it in the right column. The solutions are listed at the end of the chapter.

____	1. W_S	a.	acceptance region (α = 0.05)
____	2. T_+	b.	$\sum_i \frac{O_i^2}{E_i} - n$
____	3. $\chi^2_{.01}(4)$	c.	sum of ranks of the smaller group
____	4. $\sum_i \frac{(O_i - E_i)^2}{E_i}$	d.	degrees of freedom for a χ^2-test of two categorical variables with r and c categories, respectively
____	5. $(r - 1) \times (c - 1)$	e.	sum of positive signed ranks
____	6. $\chi^2_{obs} < \chi^2_{.05}(df)$	f.	rejection region (α = 0.05)
____	7. $\chi^2_{obs} > \chi^2_{.05}(df)$	g.	the critical value of chi-square with df = 4 for α = 0.01.

PRACTICE QUIZ

Questions

1. Susan is planning a holiday for the month of July. She wants to know whether the proportions of sunny, cloudy, and rainy days, respectively, were different last July for the four destinations she is considering. Her mother, a statistician, determines the value of the relevant test-statistic, $\chi^2_{obs} = 9.5$. Susan has been taught to use a 5% level of significance when drawing conclusions. What should she conclude?
   a) The four population distributions are equal, since $\chi^2_{obs} > 3.841$.
   b) The four population distributions are equal, since $\chi^2_{obs} < 12.59$.
   c) The four population distributions are equal, since $\chi^2_{obs} < 14.45$.
   d) The four population distributions are not equal, since $\chi^2_{obs} > 3.841$.
   e) The four population distributions are not equal, since $\chi^2_{obs} < 12.59$.
   f) The four population distributions are not equal, since $\chi^2_{obs} < 14.45$.

2. A study was conducted of the relationship between anxiety and personal space. The subjects were placed in 4-person groups, in either a very large room, a medium-sized room, a small room, or a very small room, and were asked to work on a task requiring cooperation. A questionnaire was used to measure the level of anxiety. Of the subjects in the very large room, 21 reported feeling extreme anxiety, 18 reported a high

degree of anxiety, 35 reported some anxiety, and 59 reported little or no anxiety. Of the subjects in the medium-sized room, 39 reported extreme anxiety, 47 reported a high degree of anxiety, 23 reported some anxiety, and 49 reported little or no anxiety. Of the subjects in the small room, 55 reported extreme anxiety, 49 reported a high degree of anxiety, 51 reported some anxiety, and 43 reported little or no anxiety. In the very small room, 46 reported extreme anxiety, 40 reported a high degree of anxiety, 27 reported some anxiety and 14 reported little or no anxiety.

For this test of independence between anxiety and personal space, compute the test statistic χ^2_{obs}.

a) 57.740 c) -0.666

b) 673.740 d) -600.005.

3. In a random sequence of 69 games of Black Jack in the Blue Moon Casino, a federal investigator observes 49 wins and 20 losses. Test the hypothesis that the probability of a win p_w is 0.6, as the casino advertises. What should be concluded, at the 5% level of significance?

a) $p_w = .6$ (p-value $>.10$)

b) $p_w = .6$ (p-value $<.05$)

c) $p_w = .6$ (p-value $>.05$)

d) $p_w \neq .6$ (p-value $>.10$)

e) $p_w \neq .6$ (p-value $<.05$)

f) $p_w \neq .6$ (p-value $>.05$).

4. In order to learn more about her students, a public school teacher wishes to determine whether the number of cooperative acts her students perform during the first week of September is the same as the number of cooperative acts they perform during the last week of November. The teacher keeps a record, for each student, of the number of cooperative acts performed daily during these two time periods.

   Which of the following tests would you apply?

   a) test of goodness-of-fit

   b) test of homogeneity of $\underline{J}$ independent distributions

   c) rank-sum test for two independent groups

   d) rank-sum test for two dependent groups.

5. In a review of experiments involving rats learning to run mazes it is found that the four most commonly used reinforcers are food pellets, discontinuation of electrical shock, sugar solution, and water. It is hypothesized that food pellets will be preferred in 67% of the cases, electrical shock will be preferred in 18% of the cases, sugar solution in 11% of the cases, and water in 4% of the cases. An experiment is conducted with the following results: 256 rats learned to run a maze: of these 150 learned best with food pellets, 98 with electrical shock, 6 with sugar solution, and 2 with water. In this goodness-of-fit test, compute the value of the test statistic, χ^2_{obs}.

   a) -4.875   d) 85.269

   b) -2.000   e) 251.25.

   c) -0.590

6. Marble World advertises that in every bag of 17 marbles sold, 2 are white, 4 are black, 6 are blue, 3 are green, and 2 are red. Government inspectors wish to test this claim using a 5% level of significance. After drawing a random sample of 100 marbles, the investigators compute χ^2_{obs} to be 8.839. What should they conclude?

   a) Accept $\underline{H}_o$ since $\chi^2_{obs} > 3.841$.

   b) Accept $\underline{H}_o$ since $\chi^2_{obs} < 9.488$.

   c) Reject $\underline{H}_o$ since $\chi^2_{obs} > 3.841$.

   d) Reject $\underline{H}_o$ since $\chi^2_{obs} < 9.488$.

7. While babysitting her two young cousins one evening, a statistics student observes that the two are arguing over whose frog can jump the furthest. She decides to settle the argument scientifically by using the rank-sum test of independent groups. She records the distance (in meters) jumped by each frog in the next three minutes:

David's frog	Michael's frog
2.3	4.5
1.7	3.6
4.5	3.3
4.5	2.8
6.4	5.5
9.8	4.6
9.9	4.1
	2.2

Compute the value of the test statistic:

a) 19

b) 62

c) 62.66

d) 63

e) 64.

8. 40 pigeons matched on learning ability were used in an experiment to test whether cigarette smoke hinders learning. 20 pigeons were placed in a no-smoke condition and were required to learn a task. The same was required of the pigeons in the cigarette-smoke condition. The time taken for each pigeon to learn the task was recorded and compared. In this rank-sum test for dependent groups, what would the researchers conclude given that the value of the test statistic, $\underline{T}_+$, is 159?

   a) There is a difference between the population distributions of the no-smoke and cigarette-smoke conditions ($\underline{p}$-value >.02).

   b) There is a difference between the population distributions of the no-smoke and cigarette-smoke conditions ($\underline{p}$-value <.05).

   c) Accept $\underline{H}_o$: $\underline{p}$ = .5 ($\underline{p}$-value >.02).

   d) Reject $\underline{H}_o$: $\underline{p}$ = .5 ($\underline{p}$-value <.05).

   e) Both (b) and (d) are correct.

   f) Both (a) and (c) are correct.

9. A food analyst wishes to discover whether the number of children who eat junk food is the same as the number of adults. Volunteers for the study consist of 7 children and 7 adults. A daily record is kept of the number of times junk food is eaten by each person and the value of the test statistic is computed to be 65.

   What conclusion should the analyst draw?

a) Reject H_o (p-value < .10).

b) Reject H_o (p-value > .05).

c) Reject H_o (p-value > .10).

d) Fail to reject H_o (p-value > .10).

e) Fail to reject H_o (p-value < .05).

f) Fail to reject H_o (p-value > .05).

10. In a perception experiment, the response time (in seconds) for a search task with two different stimulus arrangements are to be compared. To remove natural variability between subjects in response time, both stimulus arrangements are presented to each of seven subjects, thus permitting an analysis of the difference between stimulus arrangements within each person.

Subject	Arrangement I	Arrangement II
A	3.7	4.2
B	5.2	5.6
C	5.7	4.9
D	12.0	11.9
E	4.8	5.4
F	4.3	5.0
G	3.8	3.5

Use the appropriate rank-sum test to determine whether the two distributions of response times are the same. What is the value of the test statistic?

a) 3 c) 14

b) 10 d) 51

Analysis

Correct Answer	Explanation of Wrong Answers
1b	c this is the rejection region for $\alpha/2$ (see Study Guide Chapter 14 Tip D);
	d you used $\underline{df} = 1$ instead of $\underline{df} = (\underline{r} - 1)(\underline{c} - 1)$ for the critical value;
	e your critical value is correct, but you drew the wrong conclusion (see Text Chapter 10);
	a see d; you also drew the wrong conclusion (see Text Chapter 10);
	f see c; you also drew the wrong conclusion (see Text Chapter 10).
2a	b you forgot to subtract $\underline{n}$, the total number of subjects, in the computational formula for χ^2_{obs};
	c you forgot to square $(\underline{O}_i - \underline{E}_i)$ in the definitional formula for χ^2_{obs};
	d you forgot to square $\underline{O}_i$ in the computational formula for χ^2_{obs};
3c	f you drew the wrong conclusion given your (correct) $\underline{p}$-value (see Text Chapter 10);

Correct Answer	Explanation of Wrong Answers
	a you incorrectly multiplied the probability of the observed or a more extreme result by two when determining the *p*-value (see Study Guide Chapter 14 Tip E);
	b you used Table A8 incorrectly when determining the *p*-value (see Study Guide Chapter 14 Tip E);
	e see b; also, you drew the wrong conclusion given your *p*-value (see Text Chapter 10);
	d see a; also, you drew the wrong conclusion given your *p*-value (see Text Chapter 10);
4d	a b c see Objectives 11 and 15.
5d	e you incorrectly used p = .5 in the definitional formula for χ^2_{obs} when computing the expected values;
	c you forgot to square $(O_i - E_i)$;
	b see e and c.
	a you incorrectly used p = .5 in the computational formula for χ^2_{obs} when computing the expected values;
6b	a you used df = 1 instead of $df = (r - 1)$ for the critical value;
	d you drew the wrong conclusion given your (correct) critical value (see Text Chapter 10);
	c see a and d.

Correct Answer	Explanation of Wrong Answers
7e	a this is the value of $\underline{T}_+$, the test statistic for a rank-sum test of dependent groups; (the last distance for Michael's frog was dropped);
	b c d you should review the treatment of ties in the rank-sum test of independent groups (see Text 15.5);
8b	a you drew the wrong conclusion given your p-value (see Text Chapter 12);
	c d these are correct conclusions for a sign test ($\underline{H}_o$: $\underline{p}$ = .5), not for a rank-sum test (see Study Guide Tip B);
	e see d;
	f see a and c.
9d	c you drew the wrong conclusion given your (correct) p-value (see Text Chapter 10);
	f you reported the p-value for a one-tailed test;
	a you used Table A9 incorrectly when determining the p-value;
	b see f; you also drew the wrong conclusion for your p-value;
	e see a; you also drew the wrong conclusion for your p-value.
10b	a this is y, the number of pairs where $\underline{x}_1 > \underline{x}_2$, i.e., the test statistic for the sign test;
	c you incorrectly ranked the differences, starting with the largest difference;

Chapter 15 TESTS FOR CATEGORICAL AND RANK VARIABLES

Correct Answer	Explanation of Wrong Answers
	d this is $\underline{W}_S$, the test statistic for the rank-sum test for independent groups.

Symbol Exercise Solutions

1c, 2e, 3g, 4b, 5d, 6a, 7f.

Table SG 15.1 Example Solutions

(a) number of cups of coffee a person drinks each week;

(b) family income;

(c) race;

(d) number of children in a family;

(e) sex;

(f) eyecolor;

(g) marital status;

(h) numerical value above or below the group median;

(i) reaction times ranked;

(j) age (in years) ranked.

CHAPTER 16. CORRELATION AND OTHER MEASURES OF ASSOCIATION

OBJECTIVES

Chapter 16 continues the discussion in Chapter 8 about the relationship or association between two variables. You should be familiar with

I the general notion of measuring association;

II tests and measures of linear association (Pearson correlation);

III prediction and the Pearson correlation;

IV tests and measures of monotonic association (Spearman correlation);

V measures of general association (Cramér's measure).

More specifically, after reading the text should be able to

I K (1) define the terms *positive monotonic association*, *negative monotonic association*, *general association*, *linear association*, *curvilinear association* (Text 16.1);

(2) list two measures of association introduced in Chapter 8 (Introduction to Text Chapter 16);

(3) state the two criteria which determine the choice of the particular measure of association (Introduction to Text Chapter 16);

C (4) distinguish between tests of association and measures of association (Text 16.1);

(5) distinguish between four different types of association, their test statistics, and measures (Text 16.1; Text Table 16.1);

A (6) determine, by inspection of data represented in a graph or in a frequency table, whether an association seems to be present or absent and what type of association it is (Text 16.1):

a) association between a dichotomous and a numerical variable;

b) association between two numerical variables;

c) association between two categorical variables;

II K (7) define the term Fisher's $\underline{Z}$ (Text 16.2);

(8) state the shape, mean, and standard deviation of

a) the distribution of Fisher's $\underline{Z}$;

b) the distribution of the difference $\underline{Z}_1 - \underline{Z}_2$ (Text 16.2; Study Guide Tip A);

A (9) state $\underline{H}_o$ and $\underline{H}_1$ for the test of zero correlation (Text 16.2);

(10) carry out a test of zero correlation by determining the rejection region from Table A11 (Text 16.2);

(11) state $\underline{H}_o$ and $\underline{H}_1$ for the test of a specified (non-zero) correlation (Text 16.2);

(12) carry out a test of a specified (non-zero) correlation using Fisher's $\underline{Z}$: convert $\underline{Z}$ into the standardized normal variable $\underline{z}_{obs}$ and compare with $\underline{z}_{crit}$ (Text 16.2);

(13) determine the 95% or 99% confidence interval for ρ, based on the sample correlation $\underline{r}$, from the charts in Table A13 (Text 16.2; Study Guide Tip B);

(14) state $\underline{H}_o$ and $\underline{H}_1$ for the test of equality of two independent correlations (Text 16.2);

(15) carry out a test of equality of two independent correlations by converting the difference of two Fisher's $\underline{Z}$s ($\underline{Z}_1 - \underline{Z}_2$) into the standardized normal variable $\underline{z}_{obs}$ (Text 16.2);

III K (16) define the terms prediction function, prediction line, coefficient of determination, coefficient of alienation (Text 16.3);

(17) list four properties of the Pearson correlation (Text 16.3);

(18) state the principle of prediction (Text 16.3);

C (19) decide on the applicability of the Pearson correlation, $\underline{r}$, as a measure of association (Text 16.2; Study Guide Tip C);

(20) interpret the regression function of $\underline{y}$ on $\underline{x}$ as a prediction function of $\underline{y}$ (Text 16.3);

(21) interpret $\underline{r}^2$ as a coefficient of determination (Text 16.3);

A (22) compute the slope, $\underline{b}$, and the $\underline{y}$-intercept, $\underline{a}$, of the prediction line of $\underline{y}$ (Text 16.3);

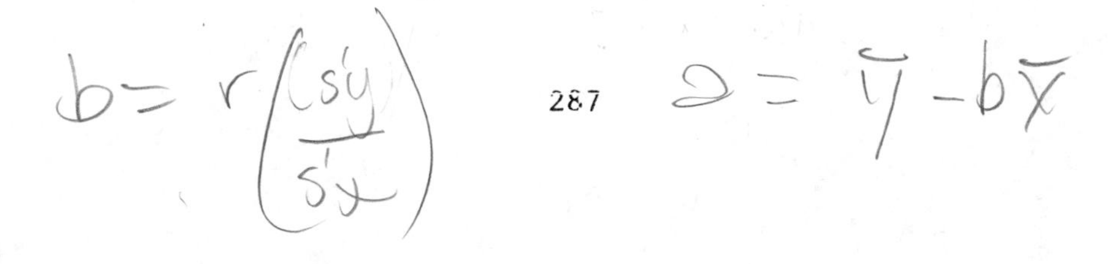

(23) determine the predicted value of $\underline{y}$ for a given value of $\underline{x}$ from the prediction function of $\underline{y}$ (Text 16.3);

IV K (24) define the term Spearman correlation (Text 16.4);

(25) state the shape, mean, and standard deviation of the distribution of $\underline{r}_S$ (Text 16.4; Study Guide Tip A);

C (26) decide on the applicability of the Spearman correlation, $\underline{r}_S$, as a measure of association (Text 16.4; Study Guide Tip C);

(27) interpret the Spearman correlation as a Pearson correlation of rank variables (Text 16.4);

A (28) compute a Spearman correlation, $\underline{r}_S$, using

a) the definitional or computational formula of the Pearson correlation on the ranked variables; or

b) the mathematically equivalent alternative formula (Text 16.4);

(29) state $\underline{H}_o$ and $\underline{H}_1$ for the test of monotonic association (Text 16.4);

(30) carry out the test of monotonic association by determining the rejection region from Table A14 (for $\underline{n} > 60$, from Table A11)

(31) compute the $100(1 - \alpha)\%$ confidence interval for ρ_S, based on the sample correlation $\underline{r}_S$ (Text 16.4);

V K (32) define the term Cramér's measure (Text 16.5);

(33) state the shape, mean, and standard deviation of the distribution of Cramér's unsquared measure, $\hat{\phi}'$ (Text 16.5; Study Guide Tip A);

(34) list the four quantities which determine the value of Cramér's measure $\hat{\phi}'^2$ (Text 16.5);

C (35) interpret the χ^2_{obs}-statistic as a statistic for the test of general association (Text 16.5);

(36) state two reasons why the χ^2_{obs}-statistic is not a useful measure of association (Text 16.5);

(37) interpret $\hat{\phi}'^2$ and $\hat{\phi}'$ (Text 16.5);

(38) compute the $100(1 - \alpha)\%$ confidence interval for ϕ' or ϕ'^2 based on the sample value $\hat{\phi}'$ or $\hat{\phi}'^2$ (Text 16.5).

TIPS AND REMINDERS

A. Distributions of Some Statistics

Table SG 16.1 gives a summary of the shape, mean, and standard deviation of various distributions introduced in Chapter 16.

B. Using Table A13 Correctly

The charts for the 95% and 99% confidence interval of ρ found in Table A13 are designed to simplify determination of these confidence intervals. Common pitfalls in using these charts can be avoided by following these simple rules (also see Text 16.2):

Table SG 16.1 Distributions of Some Statistics

Distribution of	Shape	Mean	Standard Deviation
Z	approx. normal	$Z(\rho)$, i.e. the value of Fisher's Z corresponding to the population correlation ρ	$\frac{1}{\sqrt{n-3}}$
$Z_1 - Z_2$	approx. normal	$Z_1(\rho_1) - Z_2(\rho_2)$	$\sqrt{\frac{1}{n_1 - 3} + \frac{1}{n_2 - 3}}$
r_S	approx. normal when n is large	ρ_S, the population correlation	$\frac{1}{\sqrt{n-1}}$
$\hat{\phi}'$	approx. normal when n is large	ϕ', the population value	$\frac{1}{\sqrt{nq}}$

--- Locate the value of r (i.e., the observed correlation; not ρ!) on the x-axis; place your ruler there vertically.

--- The upper limit of the wanted interval is the value of ρ at the point where the ruler intersects the upper curve for the given sample size n; correspondingly, the lower limit is the value of ρ at the point where the ruler intersects the lower curve for the given n.

--- Note that ρ ranges from -1.0 to 1.0. Do not ignore a negative sign.

C. Pearson Correlation vs. Spearman Correlation

Both the Pearson and Spearman correlation are measures of the relationship between two numerical variables (see Text Table 16.1). While Pearson's r is a measure of linear association, Spearman's r_S is a measure of monotonic association.

Note that "linear" implies monotonicity (i.e., linear = monotonic and straight), but that the opposite is not true. This introduces an asymmetry into the applicability of the Pearson and Spearman correlation: Whenever you want to test linearity (i.e., compute the Pearson correlation between two variables), the test of monotonicity (using the Spearman correlation) will give you a similar result in terms of statistical significance. The Pearson correlation will either be somewhat larger or somewhat smaller than the Spearman correlation but will always be of the same order of magnitude.

The Spearman correlation should be used when there is reason to believe that there is a nonlinear, yet monotonic, association between the two variables, or when only the subjects' ranks on the two numerical variables are known.

D. Computing the Coefficient of Determination, r^2

As discussed in Text 16.3, $r^2 = s'^2_{Pred}/s'^2_{Tot}$, i.e., the proportion of the total variance in y which can be predicted from x. However, if you have already computed the correlation, r, between x and y, the coefficient is simply obtained by squaring the value of the correlation.

If you want to verify r^2 as the ratio of the two variances, s'^2_{Pred}/s'^2_{Tot}, compute s'^2_{Unpred} in addition to s'^2_{Pred} and s'^2_{Tot} so that you can make the simple check:

$$s'^2_{Pred} + s'^2_{Unpred} = s'^2_{Tot}.$$

Such caution pays off, since the steps involved in calculating variances are a frequent source of error.

E. Symbol Exercise

To ensure your familiarity with the extensive symbolic notation introduced in Chapter 16, this Study Guide chapter has two symbol exercises. For each one, match each symbol or expression in the left column with the expression that best describes it in the right column. The solutions are listed at the end of the chapter.

Exercise I

____ 1. ρ

____ 2. Z

____ 3. $\dfrac{Z - \text{mean}(Z)}{\text{s.d.}(Z)}$

____ 4. y_{Pred}

____ 5. $r \dfrac{s'_y}{s'_x}$

____ 6. a

____ 7. $\dfrac{1}{\sqrt{n - 3}}$

____ 8. $Z(\rho_o)$

____ 9. $1 - \dfrac{6\Sigma d^2}{n(n^2 - 1)}$

____ 10. y_{Unpred}

____ 11. $\hat{\phi}'^2$

a. b

b. z_{obs}

c. Fisher's Z for the hypothesized population correlation

d. $y_{Tot} - y_{Pred}$

e. s.d. of the distribution of Z

f. the population correlation, "rho"

g. a + bx

h. Cramér's measure

i. r_S

j. $\bar{y} - b\bar{x}$

k. Fisher's "zee"

Exercise II

____ 1.	r_S	a.	the smaller of (r - 1) and (c - 1)
____ 2.	s'^2_{Tot}	b.	lower limit of the 95% confidence interval for ρ_S.
____ 3.	$H_o: \rho_S = 0$	c.	$s'^2_{Pred} + s'^2_{Unpred}$
____ 4.	$\frac{1}{\sqrt{n - 1}}$	d.	s.d. of the distribution of r_S
____ 5.	$r_S - z_{.025} \frac{1}{\sqrt{n - 1}}$	e.	y_{Tot}
____ 6.	$y_{Pred} + y_{Unpred}$	f.	s.d. of the distribution of $\hat{\phi}'$
____ 7.	$\hat{\phi}'$	g.	"phi-hat prime"
____ 8.	$\frac{1}{\sqrt{nq}}$	h.	$\hat{\phi}'^2$
____ 9.	min(r - 1, c - 1)	i.	measure of monotonic association
____ 10.	$\frac{\chi^2_{obs}}{nq}$	j.	the population Spearman correlation is zero, as a null hypothesis
____ 11.	$\bar{y}$	k.	the mean of y

Chapter 16 CORRELATION AND OTHER MEASURES OF ASSOCIATION

PRACTICE QUIZ

Questions

1. A researcher wants to establish once and for all whether there is an association between the sex of a test-taker and the mathematical aptitude score obtained. What test-statistic should she use?

    a) r

    b) r_S

    c) Z

    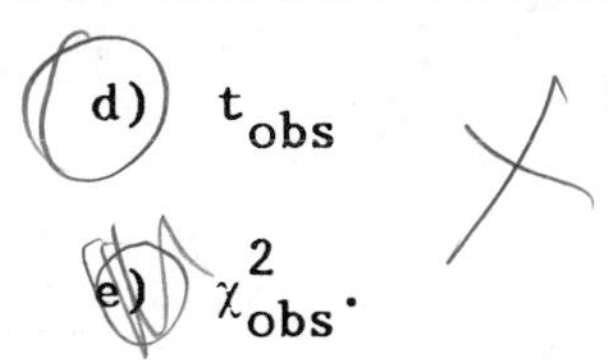

    d) t_{obs}

    e) χ^2_{obs}.

2. Does the following frequency table show evidence for an association between variables A and B? Assume the frequencies are population frequencies.

		B		
		b_1	b_2	b_3
A	a_1	20	40	60
	a_2	30	60	90

a) Yes, because the proportions of A are not equal ($\underline{a}_1 \neq \underline{a}_2$), regardless of B.

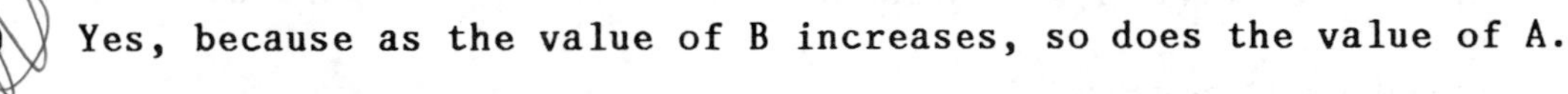

b) Yes, because as the value of B increases, so does the value of A.

c) No, because the proportions of B are spread far apart.

d) No, because the proportions of $\underline{a}_1$ and $\underline{a}_2$ are the same for all three levels of B.

3. From the following data, test whether the correlation between length of thumb and length of big toe is zero in the population.

Person	Length of Thumb (mm)	Length of Big Toe (mm)
A	31	24
B	44	27
C	40	28
D	38	25
E	34	20
F	46	35
G	37	29
H	42	31

What conclusion can you draw, based on $\alpha = 0.05$?

a) Reject H_0, since .034 < .621.

b) Reject H_0, since .782 > .621.

c) Reject H_0, since .782 > .707.

d) Fail to reject H_0, since .034 < .621.

e) Fail to reject H_0, since .034 < .707.

f) Fail to reject H_0, since .782 > .707.

4. It is thought that the population correlation between the math and science grades of high school students is 0.65. In a random sample of 100 students, a correlation of .4 was observed. Which of the following is the correct 95% confidence interval of the population correlation, based on this result?

a) $.16 < \rho < .60$

b) $.22 < \rho < .52$

c) $.22 < \rho < .55$

d) $.47 < \rho < .77$

e) $.52 < \rho < .75$

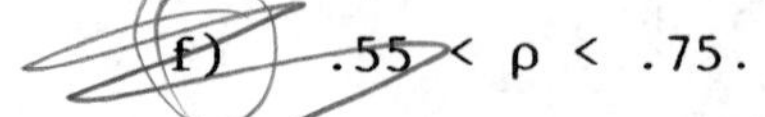

f) $.55 < \rho < .75$.

5. Test H_o: $\rho = 0.3$ against H_1: $\rho \neq 0.3$. The sample correlation is $r = 0.43$ for $n = 25$. Compute the value of the test statistic z_{obs}.

a) .354
b) .501
c) 1.661
d) .708
e) .740
f) 3.322.

6. Given the correlation $r = 0.41$ between two variables x and y and the predicted variance of y, $s'^2_{Pred} = 2.795$, what is the total variance of y?

a) 0.060
b) 0.147
c) 6.817
d) 16.627
e) insufficient information.

7. From the correlation ($r = 0.33$) between two variables x and y, their standard deviations ($s'_x = 1.26$, $s'_y = 1.56$), and their means ($\bar{x} = 8.41$, $\bar{y} = 23.2$), compute the constant term, a, and the coefficient b of the function that predicts y from x.

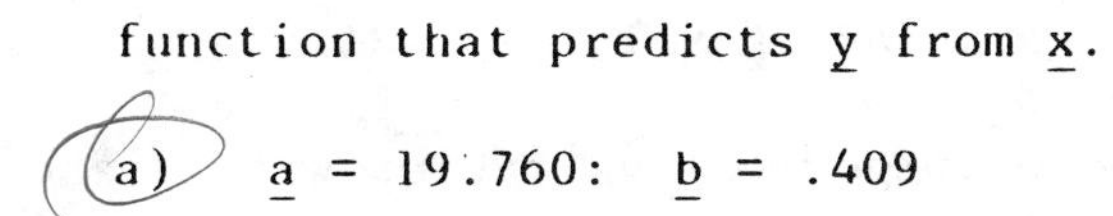

a) $a = 19.760$: $b = .409$
b) $a = 20.425$: $b = .267$
c) $a = 20.425$: $b = .409$
d) $a = 20.958$: $b = .267$.

8. Compute the Spearman rank correlation, r_S, between variables x and y for the following set of data:

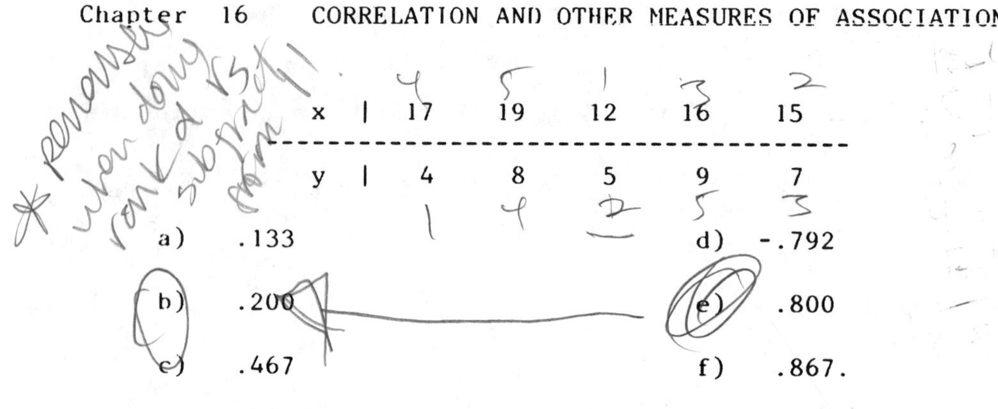
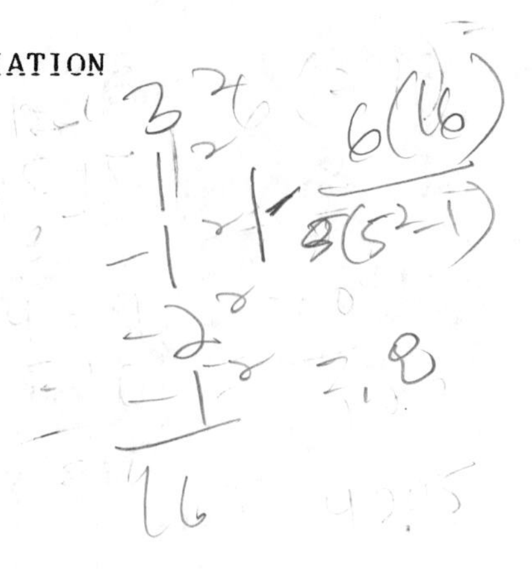

x	17	19	12	16	15
y	4	8	5	9	7

a) .133 d) -.792

b) .200 e) .800

c) .467 f) .867.

9. Test whether there is a positive monotonic association (i.e., $\underline{H}_1$: $\rho_S > 0$) between the variables $\underline{x}$ and $\underline{y}$, given that the observed Spearman correlation in a sample of size 24 was 0.512. What do you conclude?

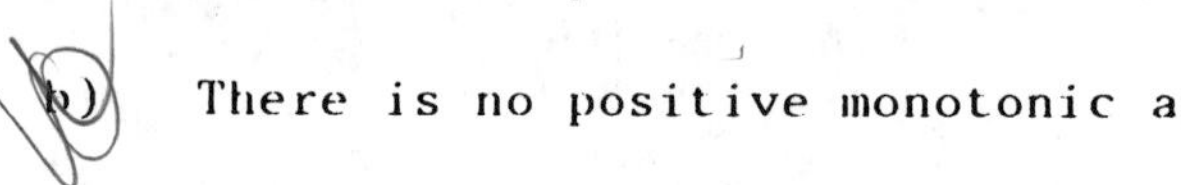

a) There is no positive monotonic association ($\underline{p}$-value < .005).

b) There is no positive monotonic association ($\underline{p}$-value < .01).

c) There is no positive monotonic association ($\underline{p}$-value < .02).

d) There is positive monotonic association ($\underline{p}$-value < .005).

e) There is positive monotonic association ($\underline{p}$-value < .01).

f) There is positive monotonic association ($\underline{p}$-value < .02).

10. In a test of association between two categorical variables with 3 and 4 categories, respectively, the test statistic, χ^2_{obs}, was 4.188. The sample size was 130. Compute the 95% confidence interval for Cramér's squared measure of association between the two variables, ϕ'^2.

a) $.00 < \phi'^2 < .06$ d) $.15 < \phi'^2 < .40$

b) $.39 < \phi'^2 < .63$ e) $.01 < \phi'^2 < .02$

c) $.01 < \phi'^2 < .25$ f) $.12 < \phi'^2 < .13$.

Analysis

Correct Answer	Explanation of Wrong Answers
1d	a b c e see Objective 5.
2d	a b equal proportions of $\underline{a}_1$ and $\underline{a}_2$ for all levels of factor B are an indication of no association;
	c the particular distribution of proportions for the different levels of one variable is irrelevant for the determination of association.
3c	f you reached the wrong conclusion given your (correct) correlation and critical value (see Text Chapter 10);
	b you forgot to split α for the two-tailed test when determining the critical value;
	e you probably used the variances instead of the standard deviations of $\underline{x}$ and $\underline{y}$ in the computation of $\underline{r}$;
	a see e and b; also you should have accepted $\underline{H}_o$ given your correlation and critical value;
	d see e and b.
4c	a you used the wrong chart (99% instead of 95% confidence interval);
	e you placed your ruler at the value of ρ instead of $\underline{r}$ (see Study Guide Tip B);
	d see a and e;
	b f see Study Guide Tip B for the correct procedure.

Correct Answer	Explanation of Wrong Answers

5d

b e you used the wrong standard deviation (s.d.) in the denominator of the formula for $\underline{z}_{obs}$ (see Table SG 16.1): in

b the s.d. of the distribution of $Z_1 - Z_2$;

e the s.d. of the distribution of $\underline{r}_s$;

a see b; in addition, you used an incorrect version of this standard deviation, taking the square root of each term and then summing them, instead of summing them before you take the square root;

f you forgot to take the square root of the s.d. of $\underline{Z}_{obs}$;

c see b and f.

6d

e the equation for the coefficient of determination,

$$r^2 = \frac{s'^2_{Pred}}{s'^2_{Tot}},$$

establishes a relationship between the three quantities $\underline{r}$ (or $\underline{r}^2$), $\underline{s}'^2_{Pred}$, and $\underline{s}'^2_{Tot}$. Knowing any two allows you to determine the third one;

a you incorrectly rearranged the equation:

$$s'^2_{Tot} = \frac{s'^2_{Pred}}{r^2}, \ \underline{\text{not}} \ \frac{r^2}{s'^2_{Pred}};$$

c you forgot to square the value of $\underline{r}$ before substituting it for $\underline{r}^2$ into the equation;

b see a and c.

Correct Answer		Explanation of Wrong Answers
7a	c	your value of $\underline{b}$ is correct; you should have used it in your computation of $\underline{a}$: $\underline{a} = \bar{\underline{y}} - b\,\bar{\underline{x}}$, not $\bar{\underline{y}} - r\,\bar{\underline{x}}$;
	d	you reversed the standard deviations of $\underline{x}$ and $\underline{y}$ when computing $\underline{b}$; given your incorrect $\underline{b}$, your computation of $\underline{a}$ is correct;
	b	your value of $\underline{a}$ is incorrect as in c; your value of $\underline{b}$ is incorrect as in d.
8b	c	the "6" in the expression $6\Sigma\underline{d}^2$ is a constant; do not use ($\underline{n}$ - 1) instead;
	d	the formula for $\underline{r}_S$ is not $\{1 - 6\Sigma\underline{d}^2\}/\{\underline{n}(\underline{n}^2 - 1)\}$;
	e	you forgot to subtract the expression $6\Sigma\underline{d}^2/\{\underline{n}(\underline{n}^2 - 1)\}$ from 1;
	f	you forgot the "6" in the expression $6\Sigma\underline{d}^2$;
	a	see e and f.
9e	b	you reached the wrong conclusion given your (correct) p-value (see Text Chapter 10);
	f	you computed the p-value for a test of a two-sided hypothesis; however, $\underline{H}_1$ is one-sided in this case;
	d	you used Table A14 incorrectly [see discussions on finding p-values in Study Guide Chapters 11 (Figure SG 11.1) and 14 (Figure SG 14.1)];
	a	see d; also, you reached the wrong conclusion given your p-value (see Text Chapter 10);

Correct Answer		Explanation of Wrong Answers
	c	see f; also, you reached the wrong conclusion given your p-value (see Text Chapter 10).
10a	c	this is the confidence interval for ϕ'; square the limits to get the confidence interval for ϕ'^2;
	e	you forgot to take the square root of nq in the limit formula $\hat{\phi}' \pm z_{\alpha/2} \cdot 1/\sqrt{nq}$;
	d	you incorrectly took the square root of nq in the computation of $\hat{\phi}'^2 = \chi^2_{obs}/(nq)$;
	b	see c and d;
	f	see c and e.

Symbol Exercise Solutions

Exercise I: 1f, 2k, 3b, 4g, 5a, 6j, 7e, 8c, 9i, 10d, 11h.

Exercise II: 1i, 2c, 3j, 4d, 5b, 6e, 7g, 8f, 9a, 10h, 11k.

CHAPTER 17. ANALYSIS OF VARIANCE

OBJECTIVES

Chapter 17 discusses a widely used statistical procedure, the analysis of variance (ANOVA). You should be familiar with

I the difference between tests of association and measures of association in the ANOVA context;

II the origin and computation of the sums of squares (SS) in the ANOVA;

III the F-test of relationship in the one-factor ANOVA;

IV measures of relationship in the ANOVA;

V the concept of interaction;

VI tests in the two-factor ANOVA.

More specifically, after reading the text you should be able to

I K (1) define the terms <u>one-factor ANOVA</u>, <u>two-factor ANOVA</u>, <u>regression function</u>, <u>coefficient of determination</u> (Text 17.1, 17.2);

(2) state of what type the independent and dependent variables are in the ANOVA (Text 17.1; Study Guide Tip A);

(3) state two complementary ways of analyzing the relationship between two variables (Text 17.2);

C (4) interpret the ANOVA as an extension of the t-test of two independent means (i.e., an independent-groups design) (Introduction to Text Chapter 17);

(5) interpret a test of relationship between two variables as a test of whether the regression function is horizontal in the population (Text 17.2);

(6) contrast the two coefficients of determination, r^2 and R^2, with respect to the regression function used to make the prediction (Text 17.2);

II K (7) define the terms group means, sum of squares (SS), total SS, SS between groups, SS within groups (Text 17.3);

(8) derive the SS formula ($SS_{Tot} = SS_{BG} + SS_{WG}$) from the prediction principle (Text 17.3);

C (9) derive the definitional formulas for SS_{Tot}, SS_{BG}, and SS_{WG} from their interpretation as total, predicted and unpredicted SS, respectively (Text 17.3);

(10) understand the mathematical equivalence of the definitional and computational formulas for the sums of squares (Text 17.3; Study Guide Tip B);

A (11) compute the three sums of squares for a one-factor ANOVA (Text 17.3; Study Guide Tip C);

III K (12) define the terms numerator degrees of freedom, denominator degrees of freedom (Text 17.5);

(13) describe the characteristics of the F-distribution (Text 17.4);

C (14) give an explanation for the particular number of degrees of freedom associated with SS_{BG} and SS_{WG} (Text 17.5);

(15) predict the effect on MS_{BG} and MS_{WG} (and thus on F_{obs}) when H_1 of the F-test is true (Text 17.5);

A (16) state H_o and H_1 of the F-test (Text 17.5);

(17) carry out an F-test:

(a) compute the value of the test statistic F_{obs} (Text 17.5);

(b) determine the rejection region for a given α (or compute the p-value of a given result) using Table A15 (Text 17.5; Study Guide Tip D);

IV C (18) contrast R^2 and ω^2 as measures of relationship between two variables in the ANOVA (Text 17.6);

(19) contrast R^2 and $\hat{\omega}^2$ as estimates of the relationship in the population (Text 17.6);

A (20) compute the coefficient of determination, R^2, from any two of the three sums of squares (Text 17.3);

(21) compute an estimate of ω^2 from a given sample (Text 17.6; Study Guide Tip E);

V K (22) define the terms *level*, *factor*, *main effect of Factor X*, *interaction* (Principle P14) (Text 17.8);

C (23) interpret the values in the cells and borders of an $r \times c$ summary table of the condition means in a two-factor ANOVA (Text 17.8);

(24) predict the effect on the interaction between factors A and B of interchanging the two factors in the regression graph (Text 17.8);

A (25) detect the presence of a main effect of Factor A or Factor B or an interaction effect (Text 17.8) from

(a) an $r \times c$ summary table of condition means;

(b) a graph of the regression function;

VI K (26) define the terms *column factor*, *row factor*, *cell* (Text 17.9);

(27) list the five *SS* that can be computed in a two-factor ANOVA (Text 17.9);

A (28) compute the five relevant *SS*, their degrees of freedom, and the mean squares (*MS*) in a two-factor ANOVA (Text 17.9; Study Guide Tip C);

(29) state H_o and H_1 for the test of a main effect (Text 17.9);

(30) carry out a test of a main effect (Text 17.9):

(a) compute the value of the test-statistic F_{obs};

(b) determine the rejection region for a given α (or compute the *p*-value of a given result) using Table A15;

(31) state H_o and H_1 for the test of interaction (Text 17.9);

(32) carry out a test of interaction (Text 17.9):

(a) compute the value of the test statistic F_{obs};

(b) determine the rejection region for a given α (or compute the p-value of a given result) using Table A15.

TIPS AND REMINDERS

A. Dependent and Independent Variables in the ANOVA

The ANOVA can only be applied when the dependent variable (i.e., the variable measured) is numerical. The independent variable in the ANOVA is the particular group or level to which a subject is assigned; i.e., the independent variable is a categorical variable, where the number of categories is equal to the number of groups (J).

B. Derivation of the Computational Formulas

(1) It is to be shown that the definitional formula

$$SS_{BG} = \sum_j n_j(\bar{y}_j - \bar{y})^2$$

is mathematically equivalent to the computational formula

$$SS_{BG} = \sum_j \frac{T_j^2}{n_j} - \frac{T^2}{N}.$$

First, we expand the definitional formula:

$$\sum_j n_j(\bar{y}_j - \bar{y})^2 = \sum_j n_j(\bar{y}_j^2 - 2\bar{y}_j\bar{y} + \bar{y}^2)$$

$$= \sum_j n_j\bar{y}_j^2 - 2\bar{y}\sum_j n_j\bar{y}_j + \bar{y}^2\sum_j n_j.$$

Since $\bar{y}_j = T_j/n_j$, and $\bar{y} = T/N$, this is equal to

$$\sum_j \frac{T_j^2}{n_j} - 2\frac{T}{N}\sum_j T_j + \frac{T^2}{N^2}\sum_j n_j .$$

Since $\sum_j T_j = T$ and $\sum_j n_j = N$, this is equal to

$$\sum_j \frac{T_j^2}{n_j} - 2\frac{T^2}{N} + \frac{T^2}{N}$$

$$= \sum_j \frac{T_j^2}{n_j} - \frac{T^2}{N} ,$$

the computational formula for SS_{BG}.

(2) Similarly, it is to be shown that

$$SS_{WG} = \Sigma(y - \bar{y}_j)^2$$

is equivalent to

$$SS_{WG} = \Sigma y^2 - \sum_j \frac{T_j^2}{n_j} .$$

For this purpose, we must extend the notation: we write y_{ij} for the ith score in the jth group. The values of i in the jth group run from 1 to n_j, the number of scores in the jth group. The summation (Σ) becomes a double summation:

$$\sum_{j=1}^{J} \sum_{i=1}^{n_j} ,$$

which we abbreviate to:

$$\sum_j \sum_i$$

The definitional formula may then be expanded as follows:

$$\sum_j\sum_i (y_{ij} - \bar{y}_j)^2 = \sum_j\sum_i (y_{ij}^2 - 2y_{ij}\bar{y}_j + \bar{y}_j^2)$$

$$= \sum_j\sum_i y_{ij}^2 - 2\sum_j\sum_i y_{ij}\bar{y}_j + \sum_j\sum_i \bar{y}_j^2 .$$

In the second term, $\bar{y}_j$ does not depend on i and can be moved outside $\sum_i$; in the third term, the sum of i is just n_j times $\bar{y}_j^2$ since the summation runs form 1 to n_j. We then have

$$\sum_j\sum_i y_{ij}^2 - 2\sum_j \bar{y}_j \sum_i y_{ij} + \sum_j n_j \bar{y}_j^2$$

$$= \sum_j\sum_i y_{ij}^2 - 2\sum_j \bar{y}_j T_j + \sum_j n_j \bar{y}_j^2 ,$$

(by the definition of $T_j = \sum_i y_{ij}$, the sum of scores in the jth group)

$$= \sum_j\sum_i y_{ij}^2 - 2\sum_j \frac{T_j}{n_j} T_j + \sum_j n_j \left(\frac{T_j}{n_j}\right)^2 ,$$

(by the definition of $\bar{y}_j = T_j/n_j$)

$$= \sum_j\sum_i y_{ij}^2 - 2\sum_j \frac{T_j^2}{n_j} + \sum_j \frac{T_j^2}{n_j}$$

$$= \sum_j\sum_i y_{ij}^2 - \sum_j \frac{T_j^2}{n_j}$$

$$= \sum y^2 - \sum_j \frac{T_j^2}{n_j} ,$$

the computational formula for SS_{WG}.

C. Computing the Sums of Squares

The following three terms are all that is needed to compute the three sums of squares for a one-factor ANOVA:

(i) Σy^2 ("the sum of y squared")

Square every score (i.e., every score in each group) and add up these squares;

(ii) $\frac{T^2}{N}$ ("T squared divided by N")

Add up every score (grand total); square this value; then divide by the total numbers of scores;

(iii) $\sum_j \frac{T_j^2}{n_j}$ ("the sum over all j of T_j squared divided by n_j")

For each group, add the scores in the group; square this value; then divide by the number of scores in the group; finally, add up all the values thus obtained for the J groups.

The first two terms [(i) and (ii)] are the same for the computation of the five sums of squares of a two-factor ANOVA. In addition, you need:

(iii) $\sum_j T^2_{.j}/(nK)$ ("the sum over all j of 'T sub dot j' squared, divided by n times K")

For each column j, add the scores in this column across all rows k; square these column totals and add them up; divide this value by the product of n, the number of scores in each cell, times K, the number of rows;

(iv) $\sum_k T^2_{k.}/(nJ)$ ("the sum over all *k* of '*T* sub *k* dot' squared, divided by *n* times *J*")

For each row *k*, add the scores in this row across all columns *j*; square these row totals and add them up; divide this value by the product of *n*, the number of scores in each cell, times *J*, the number of columns;

(v) $\sum_k\sum_j T^2_{kj}/n$ ("the sum over all *k* and *j* of '*T* sub *k j*' squared, divided by *n*")

Add the scores in each cell; square these cell totals and add them up; divide this value by *n*, the number of scores in each cell.

Note: Adding a set of scores and squaring this sum $[(\Sigma y)^2]$ is not equivalent to squaring each score and then adding up the squares (Σy^2).

D. Rejection Region in F-tests

Similar in this respect to χ^2-tests, *F*-tests of two-sided hypotheses have one-tailed rejection regions, because the *F*-statistic, like the χ^2_{obs}-statistic, is a measure of squared deviations (see Study Guide Chapter 14, Tip D). This means that α is not split into two tails when determining the rejection region: the critical value is always an upper tail percentile of the *F*-distribution, i.e., F_α.

E. Formula Translation

Steps in computing

$$\hat{\omega}^2 = \frac{(J - 1)(F_{obs} - 1)}{(J - 1)(F_{obs} - 1) + N}:$$

(1) Compute the Numerator: subtract 1 from J, the number of groups; subtract 1 from the value of the test statistic F_{obs}; multiply the two results.

(2) Compute the Denominator: subtract 1 from J; subtract 1 from F_{obs}; multiply the two results; then add N, the total number of subjects.

(3) Divide the Numerator by the Denominator.

F. Symbol Exercise

To ensure your familiarity with the extensive symbolic notation introduced in Chapter 17, this Study Guide Chapter has two symbol exercises. For each one, match each symbol or expression in the left column with the expression that best describes it in the right column. The solutions are listed at the end of the chapter.

Exercise I

____ 1.	T	a.	$\Sigma(y - \bar{y})^2$
____ 2.	SS_{Tot}	b.	predicted score for subjects in group *j*
____ 3.	R^2	c.	$\Sigma(y - \bar{y}_j)^2$
____ 4.	s'^2_{Tot}	d.	sum of scores in the *j*th group
____ 5.	$\frac{SS_{BG}}{SS_{Tot}}$	e.	$s'^2_{Pred} + s'^2_{Unpred}$
____ 6.	$\bar{y}_j$	f.	Σy
____ 7.	T_j	g.	R^2
____ 8.	SS_{BG}	h.	$\Sigma(\bar{y}_j - \bar{y})^2$
____ 9.	$\bar{y}$	i.	grand mean
____ 10.	SS_{WG}	j.	coefficient of determination

Exercise II

____ 1. MS_{BG}	a.	$\dfrac{(J - 1)(F_{obs} - 1)}{(J - 1)(F_{obs} - 1) + N}$
____ 2. ω^2	b.	$\dfrac{SS_{BG}}{J - 1}$
____ 3. df_{WG}	c.	$N - J$
____ 4. F_{obs}	d.	$\dfrac{MS_{BG}}{MS_{WG}}$
____ 5. MS_{WG}	e.	99th percentile of the F-distribution with 2 and 5 degrees of freedom
____ 6. J	f.	"omega-squared"
____ 7. $F_{.01}(2, 5)$	g.	$\dfrac{SS_{WG}}{N - J}$
____ 8. N	h.	$J - 1$
____ 9. df_{BG}	i.	total number of subjects
____ 10. $\hat{\omega}^2$	j.	number of groups

Chapter 17 ANALYSIS OF VARIANCE

PRACTICE QUIZ

Questions

1. The total variance of the dependent variable in a problem solving study is 7.53. The predicted variance is 2.11. What is the coefficient of determination?

   a) 0.280
   b) 0.389
   c) 0.720
   d) 5.420
   e) insufficient information.

2. If you obtained a value of F_{obs} equal to 0.68 in a one-factor analysis of variance, you would conclude that:

   a) there were no significant differences among the means
   b) the variances were equal
   c) the null hypothesis was rejected
   d) all of the above.

3. Fifteen rats are randomly assigned to three different conditions in order to determine which condition is more successful in producing maze learning. The results are given in the table below:

CONDITION		
Food Pellet Only	Water only	No Food or water
8.7	6.5	22.3
3.5	8.8	15.4
9.8	9.8	10.9
3.1	12.8	12.6
10.2	15.6	25.7

Compute the value of $\underline{SS}_{WG}$.

a) 261.08
b) 273.96
c) 2126.67
d) 28538.50
e) 30404.09.

4. A recent study compared 5 different types of advertising in order to determine which was most effective. Thirty subjects participated in the study. What is the rejection region for a 1% significance level?

a) $F_{obs} > 3.85$
b) $F_{obs} > 4.14$
c) $F_{obs} > 4.18$
d) $F_{obs} < 4.14$
e) $F_{obs} < 4.18$.

5. Complete the following analysis of variance table and compute the value of F_{obs}.

Source	df	SS	MS	F
Between Groups	3			
Within Groups		415.6		
Total	29	830.8		

a) F_{obs} = 0.115 d) F_{obs} = 4.831

b) F_{obs} = 0.120 e) F_{obs} = 8.325

c) F_{obs} = 0.558 f) F_{obs} = 8.658.

6. In a study comparing the durability of seven different lines of children's clothing, the following results were obtained: SS_{BG} = 123.75, df_{BG} = 6, SS_{WG} = 107.98, df_{WG} = 23. Compute the value of the coefficient of determination, R^2.

a) 0.19 d) 0.81

b) 0.47 e) 1.15

c) 0.53 f) 4.39.

7. Decide, on the basis of the information provided in Figure SG 17.1 whether there is evidence for an interaction between Factor A and Factor B.

   a) Yes, because the two lines are not parallel.

   b) Yes, because the difference in means of Factor B varies for the three levels of Factor A.

   c) No, because the two lines are not parallel.

   d) No, because the difference in means of Factor B varies for the three levels of Factor A.

   e) both a) and b).

   f) both c) and d).

Figure SG 17.1 Practice Quiz, Question 7

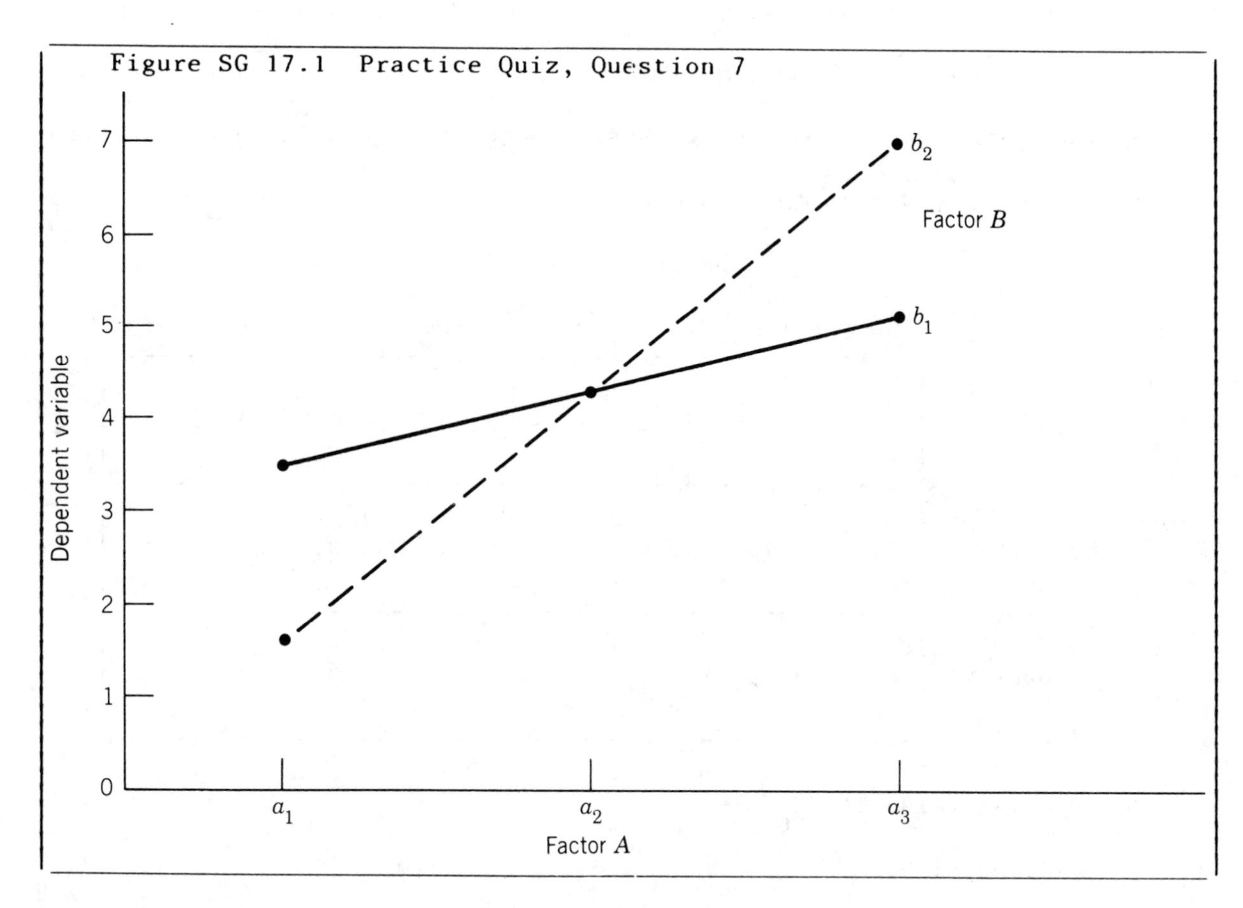

8. Twenty-four undergraduates served as subjects in a study comparing four different methods of instruction. $\underline{F}_{obs}$ was determined as 4.31. Compute the $\underline{p}$-value of this result.

a) $\underline{p}$-value > .05 since $\underline{F}_{obs}$ > 3.10

b) $\underline{p}$-value < .05 since $\underline{F}_{obs}$ > 3.10

c) $\underline{p}$-value > .05 since $\underline{F}_{obs}$ < 4.35

d) $\underline{p}$-value < .05 since $\underline{F}_{obs}$ < 4.35

e) $\underline{p}$-value > .05 since $\underline{F}_{obs}$ < 8.66.

9. An experiment was conducted to study the relationship between age and aggression under differing levels of observation. The dependent variable was the number of aggressive acts committed in each condition. The data are given in the table below:

		Observation Level (Factor A)		
		No observer	one observer	five observers
	under 25	10, 8, 7, 8, 9	3, 4, 3, 5, 5	0, 3, 1, 2, 5
Age (Factor B)	25-55	11, 12, 13, 10, 9	2, 3, 5, 4, 3	0, 2, 1, 2, 1
	over 55	8, 8, 9, 9, 12	3, 5, 3, 2, 3	0, 0, 1, 1, 3

Compute the value of $\underline{MS}_A$

a) -3961.000

b) 2.022

c) 175.570

d) 263.356

e) 526.711

f) 1056.089.

10. Complete the following ANOVA summary table and compute the F-statistic for the test of interaction.

SOURCE	SS	df	MS	F
A	22.6	4		
B	153.8	5		
A × B	25.9			
Within Groups		24		
Total	329.1			

a) 0.030
b) 0.163
c) 0.204
d) 0.245
e) insufficient information.

Analysis

Correct Answer	Explanation of Wrong Answers
1a	e see Objectives 8, 9, 20: the coefficient of determination is $R^2 = SS_{BG}/SS_{Tot} = (s'^2_{Pred}/N)/(s'^2_{Tot}/N) = s'^2_{Pred}/s'^2_{Tot}$;
	not b $s'^2_{Pred}/s'^2_{Unpred}$,
	c s'^2_{Unpred}/s'^2_{Tot}
	d s'^2_{Unpred}
2a	b c d see Objective 17b.

Correct Answer	Explanation of Wrong Answers
3a	b you computed SS_{BG} instead of SS_{WG};
	c d e you used the computational formula for SS_{WG} incorrectly (see Study Guide Tip C):
	c you computed $(\underline{T}_j/\underline{n})^2$ instead of $\underline{T}_j^2/\underline{n}$;
	d you computed $(\Sigma y)^2$ instead of $\Sigma \underline{y}^2$;
	e both c and d.
4c	a b you used the wrong degrees of freedom for the critical value:
	a $\underline{df}_1$ should be $\underline{J} - 1$, not J;
	b $\underline{df}_2$ should be N - J, not N - (J - 1);
	e you selected the correct critical value but located the rejection region on its wrong side; see Objective 17b;
	d see b and e.
5f	a c d you computed the wrong ratio of mean squares:
	$F = MS_{BG}/MS_{WG}$, not
	a MS_{WG}/MS_{BG},
	c MS_{WG}/MS_{Tot},
	d MS_{BG}/MS_{Tot};
	e since $df_{Tot} = N - 1 = 29$ and $df_{BG} = J - 1 = 3$, df_{WG} should be 26, not 25;
	b see a and e.

Correct Answer	Explanation of Wrong Answers
6c	a b d e f see Objective 20: $R^2 = SS_{BG}/SS_{Tot}$, not
	b SS_{WG}/SS_{Tot},
	e SS_{BG}/SS_{WG},
	d MS_{BG}/MS_{Tot},
	a MS_{WG}/MS_{Tot},
	f MS_{BG}/MS_{WG}.
7e	a b the reason for interaction given in answer b implies that the two lines are not parallel (answer a); i. e., either one is evidence for interaction:
	c d f see Objective 25b.
8b	c e you used the wrong degrees of freedom when looking up the critical value in the F-table:
	c df_1 should be J - 1, not 1;
	e you reversed df_1 = J - 1 and df_2 = N - J;
	a d you should review the discussion of p-value in Text and Study Guide Chapters 11 and 14; you used the wrong inequality sign given your observed and critical value; in a the critical value is correct; for d see c.
9d	b you computed MS_B instead of MS_A;
	e you computed SS_A; to obtain MS_A, divide SS_A by its degrees of freedom;

Correct Answer	Explanation of Wrong Answers
	c you divided SS_A by the wrong degrees of freedom (J instead of J − 1);
	a f you used the computational formula for SS_A incorrectly (see Study Guide Tip C):
	a you computed $\underline{T}^2/\underline{n}$ instead of $\underline{T}^2/\underline{N}$;
	f you computed $(\Sigma \underline{T}_{.j})^2/(\underline{nK})$ instead of $\Sigma \underline{T}^2_{.j}/(\underline{nK})$;
10d	a you computed the wrong ratio of mean squares as $\underline{F}$ (see Objective 32a);
	b you used the wrong degrees of freedom: $\underline{df}_{AB}$ = (J − 1)(K − 1), not JK;
	c you computed $\underline{F}$ as a ratio of sums of squares instead of mean squares;
	e all relevant information--$\underline{SS}_{WG}$ and $\underline{df}_{AB}$--can be determined.

Symbol Exercise Solutions

Exercise I: 1f, 2a, 3j, 4e, 5g, 6b, 7d, 8h, 9i, 10c.

Exercise II: 1b, 2f, 3c, 4d, 5g, 6j, 7e, 8i, 9h, 10a.

REVIEW CHAPTER D

OBJECTIVES

The various statistical procedures described in Chapters 1 to 17 are reviewed and classified in this chapter. You should be familiar with

I the different types of variables;

II the classification of statistical procedures;

III the requirements for the validity of statistical procedures;

IV nonparametric and distribution-free procedures.

More specifically, after reading the text you should be able to

I K (1) define and give examples of the four principal types of variables (Text D.1; see also Text 15.1 and Study Guide Chapter 15, Objectives 1 and 2);

II A (2) select the appropriate procedure for a test or estimate (Text D.2, D.3);

III K (3) list and describe the six requirements for the validity of statistical procedures (Text D.4);

C (4) decide whether a given requirement applies to a particular procedure (Text D.4);

IV K (5) define the terms <u>nonparametric test</u> and <u>distribution-free test</u> (Text D.5);

(6) give examples of nonparametric and distribution-free procedures (Text D.5);

C (7) describe two situations in which nonparametric and distribution-free procedures are used (Text D.5).

CHAPTER 18. ADVANCED PROBABILITY

OBJECTIVES

Chapter 18 discusses probability theory which is the foundation of statistics. You should be familiar with

I the product rule and its special cases which allow you to compute the number of ordered and unordered samples in a sample space;

II the notion of events and their probabilities;

III the independence and dependence of events.

More specifically, after reading the text you should be able to

I K (1) define the terms product rule, ordered samples, permutations, unordered samples, combinations, element, sample point, binomial coefficient (Text 18.1, 18.2, 18.3, 18.4);

C (2) explain why sampling without replacement results in fewer sample points than sampling with replacement (Text 18.1);

(3) explain why there are fewer sample points for unordered than for ordered samples (Text 18.2);

(4) interpret the formulas to compute the number of ordered and unordered samples of size n selected from a population of size N with or without replacement as special cases of the product rule (Text 18.3);

(5) interpret the binomial coefficient, i.e., the number of paths with y successes in n trials, as a special case of the formula for the number of subsets of size n from a population of size N (Text 18.4);

A (6) represent the sample spaces of the following experiments by a tree diagram (Text 18.1):

a) taking ordered samples of size n from a population of size N with replacement;

b) taking ordered samples of size n from a population of size N without replacement;

c) arranging a set of size N in all possible ways;

(7) list the unordered samples of size n selected from a population of size N (Text 18.2; Study Guide Tip B);

(8) compute the number of unordered samples of size n given the number of ordered samples and the sample size (Text 18.2);

(9) compute the number of

a) ordered samples of size n taken from a population of size N with replacement (Text 18.1);

b) ordered samples of size n taken from a population with large N without replacement (Text 18.1);

c) ordered samples of size n taken from a population with small N without replacement (Text 18.1);

d) ways to arrange a set of size N (Text 18.1);

e) unordered samples of size n taken from a population of size N without replacement (Text 18.2);

f) possible subsets of size n of a set of size N (Text 18.2);

g) samples of size n with a given number of choices at each stage (Text 18.3).

(10) compute the number of sample points with a specified characteristic from a given population with help of the product rule (Text 18.3);

II K (11) define the terms stage, event, stage event, elementary event, independent stages/trials, dependent stages/trials, union, intersection, complement, Venn diagram, mutually exclusive events (Text 18.5, 18.6);

C (12) interpret the union, intersection, and complement of events as events defined in terms of other events (instead of in terms of sample points) (Text 18.5);

(13) distinguish between the independence of trials/stages and the fairness or bias of a coin or die (Text 18.6);

A (14) decide (with help of a tree diagram) whether two stages are independent or dependent (Text 18.6; Study Guide Tip C);

(15) decide whether two events are mutually exclusive or not (Text 18.5);

(16) represent the sample space of an experiment by a matrix (for two-stage experiments) or by a tree diagram (Text 18.5);

(17) represent the union, intersection, and complement of given events by a Venn diagram (Text 18.5);

(18) compute the probability of a specified event for sample spaces where

a) all sample points have equal probability (Text 18.5);

b) different sample points have different probabilities (Text 18.5);

(19) compute the probability of the complement of an event given the probability of the event (Text 18.5);

(20) compute the probability of the union of two events if

a) the two events are mutually exclusive (Text 18.5);

b) the two events are not mutually exclusive (Text 18.5);

III K (21) define the terms intersection probability, unconditional probability, conditional probability, conditional event, independent events, dependent events (Text 18.7);

C (22) interpret a path through a tree diagram as the intersection of stage events (Text 18.7);

(23) distinguish between the independence of stages and the independence of events (Text 18.7);

(24) decide whether two events are dependent or independent (Text 18.7);

(25) distinguish between independent events and mutually exclusive events (Text 18.7);

(26) distinguish between the conditional events A|B and B|A (Text 18.7);

A (27) compute a conditional probability for a two-stage tree diagram given the relevant intersection probability and unconditional probability (Text 18.7);

(28) compute the intersection probability of two independent events (Text 18.7);

TIPS AND REMINDERS

A. Computing the number of sample points

Figure SG 18.1 summarizes the different formulas for computing the number of samples of size n which can be selected from a population of size N.

Figure SG 18.1 Computing the Number of Sample Points

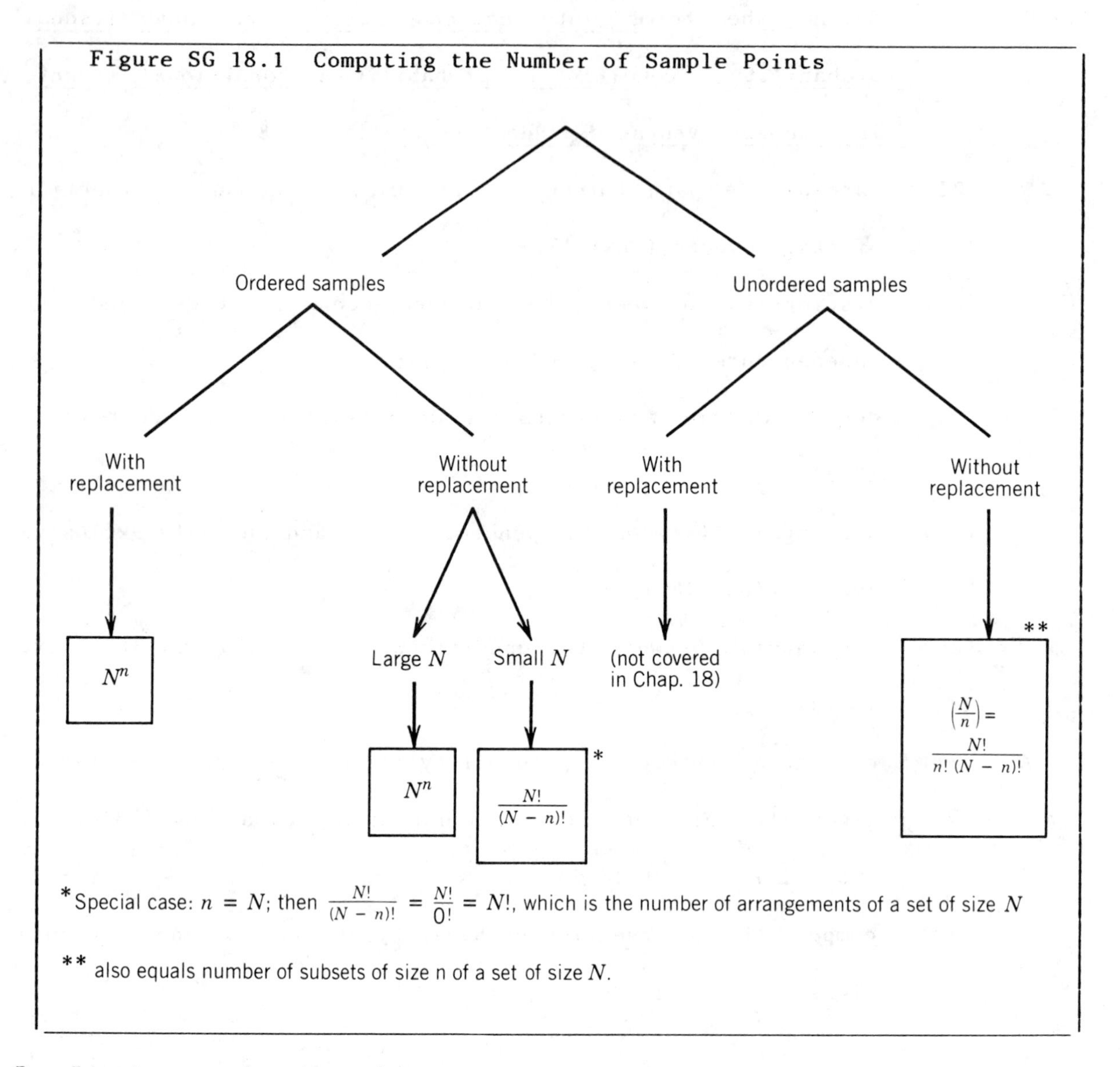

*Special case: $n = N$; then $\frac{N!}{(N-n)!} = \frac{N!}{0!} = N!$, which is the number of arrangements of a set of size N

** also equals number of subsets of size n of a set of size N.

B. Listing unordered samples

It is difficult to represent the unordered samples of size n which can be selected without replacement from a population of size N by a tree diagram. Instead, the sample points are usually simply listed. Listing them in a systematic way prevents you from leaving out any sample point. The following two examples illustrate one such way to list samples.

Example 1

Population: A D F G K

Task: List all possible unordered samples (without replacement) of size 2.

Procedure:

a) Take the first population element and combine it with all population elements that follow it:

AD AF AG AK

b) Take the second population element and combine it with all population elements that follow it:

DF DG DK

(Do not combine it with the elements that precede it, i.e., DA. DA is the same as AD which was listed in step a.)

c) Take the third population element and combine it with all population elements that follow it:

FG FK

d) Continue step c (taking the fourth, fifth, and so on element) until you obtain only one sample point, in this case: GK

Solution: The list of all possible unordered samples of size 2 is: AD AF AG AK DF DG DK FG FK GK.

Example 2

Population: 2 4 5 7 8 9

Task: List all possible unordered samples (without replacement) of size 3.

Procedure:

a) Take the first and second population element and combine them with all population elements that follow the second element:

245 247 248 249

b) Take the first and the third element and combine them with all elements that follow the third element:

257 258 259

c) Take the first and the fourth (fifth and so on) element and combine them with all elements that follow the fourth (fifth and so on) element until you obtain only one sample point (i.e., 289):

278 279

289

d) Now start with the second element: Take the second and third element and combine them with all elements that follow the third element:

457 458 459

e) Take the second and fourth (fifth and so on) element and combine them with all elements that follow the fourth (fifth and so on) element until you obtain only one sample point (i.e., 489):

478 479

489

f) Now start with the third element: Take the third and fourth element and combine them with all elements that follow the fourth element; repeat steps d - f until you obtain only one sample point (i.e., 789):

578 579

589

789

Solution: The list of all possible unordered samples of size 3 is:

245 247 248 249 257 258 259 278 279 289 457

458 459 478 479 489 578 579 589 789

C. Dependent and independent trials/stages

A sequence of trials is defined as independent if, for each trial, the subtrees in the tree diagram of the sample space are all identical. Conversely, a sequence of trials is dependent if, for any trial, the subtrees are not all identical.

The subtrees in a tree diagram can differ in several ways:

a) in the labels of the branches (i.e., the stage events that the branches represent),

b) in their number of branches,

c) in the probability of each branch,

and all combinations of a, b, and c. If the subtrees for any one trial differ in any of those ways, the sequence of trials or stages is dependent.

D. Symbol Exercise

Match each symbol or equation in the left column with the expression in the right column that best describes it. The solutions are listed at the end of the chapter.

_____	1.	N	a.	$P(A) + P(B) - P(AB)$
_____	2.	n	b.	the probability of B given A.
_____	3.	$\binom{N}{n}$	c.	$P(\bar{A})$
_____	4.	AB	d.	the size of a sample.
_____	5.	$A \cup B$	e.	$\frac{N!}{n!\,(N - n)!}$
_____	6.	$\bar{A}$	f.	the union of events A and B.
_____	7.	$P(A \cup B)$	g.	$P(AB)$
_____	8.	$P(B\|A)$	h.	the probability of both A and B occurring.
_____	9.	$P(AB)$	i.	the number of elements in a population.
_____	10.	$1.0 - P(A)$	j.	the intersection of events A and B.
_____	11.	$P(A) \times P(B\|A)$	k.	the complement of event A.

PRACTICE QUIZ

Questions

1. An employer can fill three entirely different positions in 504 different ways from the pool of applicants. If all three positions were identical, in how many different ways could they be filled by the same number of applicants?

   a) 84

   b) 168

   c) 1512

   d) 3024

   e) impossible to say without knowing N, the number of applicants.

2. A manufacturer assembles a car by choosing an engine size, a type of transmission, and a body style. How many different cars can possibly be manufactured if there are 5 sizes of engines, 4 types of transmission, and 7 body styles?

   a) 4096

   b) 3360

   c) 560

   d) 140.

3. An employer interviews eight persons for possible employment. Two persons are to be hired. Five of the eight interviewed persons are women, three are men. In how many different ways could the employer select two women?

   a) 10

   b) 20

   c) 25

   d) 56

   e) 336.

4. Random samples of 7 boys are drawn without replacement from a population of 18 boys, of whom 11 are Catholic and 7 are Protestant. Of all the different samples that can be drawn, how many have exactly 4 Catholic and 3 Protestant boys?

   a) 1,663,200
   b) 11,550
   c) 8,130
   d) 365
   e) 144.

5. A die is rolled. It is observed whether the number that appears is odd or even (first stage). If it is odd, the actual number is recorded (second stage); if it is even, the square of the actual number is recorded (second stage). Are the two stages of this experiment dependent or independent?

   a) dependent, because the subtrees in the tree diagram representing the experiment are different for the two stages.

   b) independent, because there is an equal probability of obtaining an odd or an even number at the first stage.

   c) dependent, because the subtrees in the tree diagram representing the experiment are different at the second stage depending on the result at the first stage.

   d) independent, because the subtrees in the tree diagram representing the experiment have an equal number of branches at the second stage.

6. Two tickets are drawn at random from an envelope containing seven tickets. Two of the tickets are green, three are red, and two are blue. C is defined as the event that at least one of the two tickets is green. D is defined as the event that both tickets are of the same color. Compute $P(C \cup D)$.

   a) 5/7 d) 32/49

   b) 16/21 e) 37/49

   c) 30/49 f) 41/49.

7. Four types of soft drink (cola, rootbeer, orange, and gingerale) are judged for thirst-quenching ability by a subject in an experiment. The drink he thinks is best is ranked one, the next best is ranked two, etc. The subject, in fact, finds that all the drinks are equally thirst-quenching so that any order of the ranking is equally likely. What is the probability that cola is ranked either first or second?

   a) 1/16 d) 7/12

   b) 1/12 e) 1/2.

   c) 3/8

8. Two fair dice are tossed. P is defined as the event that the number showing in the first die is 4 or less. Q is defined as the event that the number showing on the second die is at least 3. Compute the probability that both P and Q will occur.

a) 1/9
b) 4/9
c) 4/27
d) 13/18
e) 8/9.

9. A company, as part of a sales promotion for one of its products, has placed coupons in half the boxes in which the product is sold. The coupon has a number printed on it; certain of these numbers are winning numbers for which a prize is awarded. Overall, there is one chance in ten of winning a prize if you buy a box. What is probability of winning a prize if you get a coupon?

a) 0.05
b) 0.1
c) 0.2
d) 0.5.

10. Two marbles are drawn in the following way. One marble is drawn at random from Bag #1 which contains blue and green marbles in the proportions: blue = ½, green = ½. The other marble is drawn at random from Bag #2 which contains yellow and red marbles in the proportions: yellow = ¼, red = 3/4. What is the probability of drawing a blue marble or a red marble or both?

a) 1/4
b) 3/8
c) 5/4
d) 7/8.

Analysis

Correct Answer	Explanation of Wrong Answers
1a	b you should have divided the number of permutations by $n!$ (not by n) (see Objective 8: the number of unordered samples is equal to the number of ordered samples divided by $n!$);
	d you should have divided (not multiplied) the number of permutations by $n!$;
	c see answers b and d;
	e see Objective 8.
2d	a b c you should have used a tree diagram and/or applied the product rule (see Objective 9g);
3a	b you should have computed the number of unordered samples;
	c you should have computed the number of unordered samples without replacement;
	d you used the correct formula but computed the number of samples of $n = 5$ taken from $N = 8$ (you should compute the number of samples of $n = 2$ taken from $N = 5$);
	e see answers b and d;
4b	a you correctly applied the product rule, but should have computed the product of the number of unordered samples;
	d you should have applied the product rule (see Text 18.3);
	c see answers a and d;

Correct Answer	Explanation of Wrong Answers
	e see Text 18.3; you computed the number of different ways 4 Catholics and 3 Protestants can be arranged.
5c	a b d see Objective 14 and Study Guide Tip C.
6a	b you computed $P(A \cup B)$ as if the two events were mutually exclusive; yet, $P(A \cup B) = P(A) + P(B) - P(AB)$;
	c the branch probabilities at the second stage of your tree diagram are probably incorrect: after one ticket is chosen, the denominator of the probabilities at the second stage decreases by 1;
	d see answers b and c;
	e you computed $P(A \cup B)$ as if sampling was with replacement;
	f see answers b and e.
7e	a you should have added the probabilities of the two mutually exclusive events (i.e. that cola is first or second);
	c you probably counted the number of paths in the tree diagram where cola is either first or second incorrectly;
	d you should have added the probabilities of all paths where cola is first or second, not just the probabilities of one branch;
	b see answers a and d.

Correct Answer	Explanation of Wrong Answers
8b	a you misunderstood the definition of event Q: second die showing a number of 3 or more, not exactly 3;
	e you should have computed the intersection probability of the two independent events P and Q, not the probability of their union;
	d see answers a and e;
	c you computed P(PQ) as if P and Q were dependent events.
9c	a b d you should have computed the conditional probability of winning a prize, given a coupon is found in the box, i.e., $P(W\|C) = \frac{P(WC)}{P(C)}$, not P(WC) × P(C) (answer a); not P(WC), the probability of winning a prize (answer b); not P(C), the probability of getting a coupon (answer d).
10d	c you conputed the union of two events as if the two events were mutually exclusive (see answer 6b);
	b you computed the intersection of the two events;
	a this is the probability of drawing a yellow marble.

Symbol Exercise Solutions

1i, 2d, 3e, 4j, 5f, 6k, 7a, 8b, 9h, 10c, 11g.